Data-driven Optimization and Control for Autonomous Energy Systems

Gang Wang · Jian Sun · Jie Chen

Data-driven Optimization and Control for Autonomous Energy Systems

北京理工大学出版社
BEIJING INSTITUTE OF TECHNOLOGY PRESS

Springer

Gang Wang
School of Automation
Beijing Institute of Technology
Beijing, China

Jian Sun
School of Automation
Beijing Institute of Technology
Beijing, China

Jie Chen
Harbin Institute of Technology
Harbin, China

ISBN 978-981-95-1781-7 ISBN 978-981-95-1782-4 (eBook)
https://doi.org/10.1007/978-981-95-1782-4

Jointly published with Beijing Institute of Technology Press
The print edition is not for sale in China (Mainland). Customers from China (Mainland) please order the print book from: Beijing Institute of Technology Press.

This Springer imprint is published by the registered company Springer Nature Singapore Pte Ltd.
The registered company address is: 152 Beach Road, #21-01/04 Gateway East, Singapore 189721, Singapore

Contents

Chapter 1
Introduction

1.1 Grid Monitoring

As the most monumental engineering achievement of the twentieth century, the North American power grid stands as a complex cyber-physical system, boasting thousands of miles of transmission lines and millions of miles of distribution lines. This intricate network interconnects thousands of power plants with millions of factories and residences. The contemporary power grid faces significant challenges due to unprecedented load peaks and voltage fluctuations stemming from the continual growth of distributed renewable energy sources, electric vehicles, and demand response programs. Given the above description, it becomes exceptionally vital to delve into the real-time monitoring of smart grids. This endeavor not only pertains to the detection and protection against system instabilities (Zhang et al. 2019a; Salehi et al. 2012; Giannakis et al. 2013), but also plays a pivotal role in energy management (Zhang et al. 2016; Rathor and Saxena 2020; Koutsopoulos and Tassiulas 2011).

On the one side, the intersection of distributed generation (DG), electric vehicles, and resilient loads is introducing intricate and dynamic operational cases to electricity distribution networks (Turitsyn et al. 2011). In the United States, several public utilities are grappling with the formidable task of seamlessly integrating residential and commercial solar generation into the grid. More specifically, solar farms are typically situated in remote rural areas near the terminus of lengthy distribution lines, which can give rise to voltage regulation difficulties. Moreover, frequent reverse current flows exert significant pressure on the transformer capacity of substations. Additionally, data gathered from residential photovoltaic (PV) systems reveals that solar power generation can fluctuate by as much as 15% of its rated power within a single minute (Barker et al. 2012). All of these challenges converge to constitute a substantial obstacle in the realm of energy management within power distribution networks.

G. Wang et al., *Data-driven Optimization and Control for Autonomous Energy Systems*,
https://doi.org/10.1007/978-981-95-1782-4_1

On the flip side, modern DG units come equipped with what are known as smart inverters. These smart inverters boast bi-directional communication and computational capabilities, thereby opening up unprecedented opportunities for control (Schauder 2014). Leveraging these smart inverters, it is possible to engage in both reactive power compensation and active power reduction to attain various goals, including the minimization of power losses, the preservation of voltage stability, or the execution of voltage regulation. Historically, power distribution networks have leaned on utility equipment like load tap transformers, capacitor banks, and voltage regulators for achieving voltage regulation (Kersting 2018) Nevertheless, the operation of these devices is typically discrete, and frequent switching actions can curtail their lifespan (Schauder 2014; Agalgaonkar et al. 2015). As the prevalence of distributed renewable energy generation continues to surge, voltage regulation will necessitate more frequent switching actions and potentially additional equipment.

Monitoring and controlling the operational status of the grid has become an imperative research element, and techniques such as optimization and deep neural network models can be implemented to improve the accuracy and efficiency of grid monitoring. Specifically, optimization techniques are implemented on the results of grid state identification to locate the optimal grid configuration and operation strategies to maximise efficiency and security (Babaeinejadsarookolaee et al. 2019; Riffonneau et al. 2011) Alternatively, wisely combining stochasticity with system flexibility, stochastic optimization problems are investigated to improve the efficiency of grid operation (Bazrafshan and Gatsis 2016). Additionally, the power grid consists of thousands of devices, each of which may have an implication on the security of the system. When trained employing data collected from a small set of system variables, neural networks (NNs) have shown their capability to quickly and accurately predict system security, gaining the attention of a widening number of scholars (Wang et al. 2021; Menke et al. 2019).

1.2 State Estimation

In the 1960s, power system engineers began exploring voltage monitoring at critical buses using data collected from current and potential transformers to comprehensively monitor the grid's operational status. However, the application of power flow equations faced practical challenges due to timing and modeling inaccuracies. A pioneering investigation in (Schweppe and Wildes 1970) laid the statistical foundations for power system state estimation (PSSE). The primary challenge of PSSE is to deduce voltage magnitude and angle information for all buses, given known network parameters and measurements obtained from the entire grid.

Frequently employed criteria in PSSE encompass the weighted least squares (WLS) and the least-absolute value (LAV) Caro and Conejo (2012) For enhanced robustness, alternative estimators include the Schweppe-Huber Generalized M-Estimator (Mili et al. 1996b; Zhao and Mili 2018; Zhao et al. 2018), in addition to the Least-Median and the Least-Trimmed Squares state estimators (Mili et al.

1996b). The subsequent section primarily focuses on elaborating the WLS and LAV methods.

The WLS criterion is consistent with the maximum likelihood criterion assuming the presence of additive Gaussian white noise. Unfortunately, the WLS approach suffers from several limitations. Primarily, WLS estimation based on supervisory control and data acquisition (SCADA) measurements is tantamount to minimizing non-convex quadratic polynomials. Consequently, the obtained iterative estimation results depend significantly on the setting of the initial values (see the arguments in (Wang et al. 2017 and 2018a for details). Moreover, for non-convex optimization problems, it is typically difficult to guarantee the convergence of the Gaussian Newton iterations (Bertsekas 1999). Importantly as well, WLS estimation methods are remarkably sensitive to outliers (Merrill and Schweppe 1971). When outliers are presented, they can lead to estimates that are significantly biased. Methods to address these issues involve excluding undesirable data using the maximum normalised residual largest normalized residual (LNR) test (Merrill and Schweppe, 1971) or by reformulating the (potentially regularised) WLS into semi-deterministic programming semi-definite program (SDP) form (Zhu and Giannakis 2014; Zhang et al. 2017). The former is performed by alternating LNR tests and estimation, while the latter involves the solution of the SDP problem. Under certain conditions, minimum median squares and minimum trimmed squares estimation methods have been demonstrated to deliver improved performance (Mili et al. 1994). However, unfortunately, their computational complexity and storage requirements increase with the number of buses in the network (Zhang et al. 2019a).

LAV estimators accurately estimate the state while being able to identify and exclude erroneous data (Kotiuga and Vidyasagar 1982). Recently, investigations have concentrated on addressing the non-convexity and non-smoothness issues in LAV estimation. This includes linearizing the nonlinear measurement function in the latest iteration and tackling a range of linear programming problems (Kotiuga and Vidyasagar 1982). Improvements in linear programming techniques have also been reported by capitalizing on the structure of the system (Abur and Celik 1991) or employing iterative reweighting methods (Jabr and Pal 2003). Research has also examined LAV estimation based solely on Power Management Unit (PMU) data (Göl and Abur 2014; Xu and Abur 2017) and recommended strategic scaling methods to mitigate the impact of leverage measurements (Mili et al. 1996b, 1994). However, despite these efforts, LAV estimators are still not extensively implemented in power networks today, primarily on account of their computational inefficiency (Göl and Abur 2014).

In this work, LAV-based PSSE is revisited from the perspective of composite optimization (Burke 1985; Lewis and Wright 2016). Within this framework, the minimization function $f(v) = h(c(v))$ is taken into account, which consists of a convex function h and a smooth vector function $c(v)$.. Two new near-linear (prox-linear) methods are presented by tackling a sequence of convex quadratic subproblems. The first method is a deterministic LAV solver that operates by minimizing a function constructed from a linearised approximation of the original objective function and a quadratic regularisation. Each subproblem can be efficiently addressed either by alternating direction method of multipliers (ADMM) or by employing existing solvers and multipliers. The convergence of this deterministic near-linear algorithm has been demonstrated (Lewis and Wright 2016; Davis and Drusvyatskiy 2019). The second LAV solver is built upon the foundation of a stochastic prox-linear algorithm, with per iteration aimed at minimizing the sum of linearized approximations of individual measurement LAV losses and regularization terms (Lewis and Wright 2016; Davis and Drusvyatskiy 2019). Intriguingly, what sets the stochastic LAV solver apart is that per iteration can be accomplished through closed-form updates. Leveraging the inherent sparse connectivity of the power grid, this implies that only a minority of elements in the state vector need to be updated. Furthermore, by intelligently processing measurements in small batches, the stochastic solver can achieve faster performance.

1.3 Data-Driven

To effectively navigate the dynamic and unpredictable uncertainties inherent in smart grid configurations, leveraging historical data for state estimation (SE) is strongly recommended. There are two key rationales behind this recommendation. Firstly, the exponential advancement of cutting-edge sensing technologies, data storage capacities, and computational prowess, coupled with targeted investments in advanced instrumentation and synchronized phasor technologies, have become pivotal drivers and sources of unprecedented data for the power industry. Secondly, the quality of historical data can be further optimized through the synergistic application of human–computer interaction, power system physics principles, pattern recognition, and time series analysis. For instance, by integrating system update reports, it becomes possible to eliminate topological errors and identify and correct bad data with absolute precision. This reevaluated historical state will also yield a heightened level of accuracy.

As a result, a vast array of resources derived from historical data can be effectively harnessed to optimize the operation of power systems. In practical terms, data-driven methodologies have found extensive application across a spectrum of domains within power system analysis, including fault detection (Jin et al. 2010; Kruger et al. 2013; Yin et al. 2014, SE Weng et al. 2013), topology estimation (Weng et al. 2014; Liao et al. 2015), bad data identification (Uddin et al. 2015), wind turbine model recognition (Veen et al. 2013), battery estimation (Xiong et al. 2014), system risk assessment (Adeodato et al. 2010), wind speed forecasting (Kusiak and Zhang 2010), and ice load estimation (Li et al. 2011), among others. When it comes to PSSE, dynamic predictions can be achieved through time series analysis of historical data (Da Silva et al. 1983; Morvaj 1985; Hassanzadeh and Evrenosoğlu 2012), employing techniques such as the autoregressive moving average model method and Kalman filtering (Hassanzadeh and Evrenosoğlu 2012; Durgaprasad and Thakur 1998). For added robustness, the incorporation of anomaly detection and identification can also be considered (Durgaprasad and Thakur 1998; Nishiya et al. 1982).

1.4 Organization

Over the next seven chapters, we will sequentially introduce power system models for SE and data-driven approaches. Here's an overview of the content covered in each chapter.

Chapter 2 introduces two innovative algorithms for efficient least-absolute-value estimation. These algorithms leverage recent advancements in composite optimization, offering a substantial reduction in computation times for medium to large-sized networks when compared to existing methods.

Chapter 3 suggests a simple algebraic technique to transform the power flows into rank-one measurements, for which the ℓ_1-based misfit is minimized and develops conditions on the ℓ_1-based loss function such that exact recovery and quadratic convergence of the proposed scheme are guaranteed.

Chapter 4 introduces a novel model-specific deep neural networks (DNNs) tailored for real-time PSSE. This network is constructed by unrolling an iterative solver initially designed with an exact AC model, making it capable of delivering real-time results with only offline training and minimal tuning effort.

Chapter 5 delves into the data-driven priors problem, building upon the advancements introduced in Chap. 4. We unroll an alternating minimization solver to construct the Gauss–Newton unrolled neural networks (GNU-NNs) with deep priors.

Chapter 6 presents an ergodic energy management (EEM) framework and explores a stochastic optimization problem within a distribution grid, leveraging the capabilities of a stochastic approximation solver to effectively address the challenges involved.

Chapter 7 extends the stochastic optimal control scheme introduced in Chap. 6. It combines data-driven and physical optimization principles to introduce a novel two-timescale voltage regulator specifically designed for distribution networks.

Chapter 8 introduces an unsupervised statistical learning approach designed for computationally intensive and time-sensitive reactive power control. Additionally, it presents the deep policy gradient approach, aimed at enhancing both the computational efficiency and robustness of this control method when dealing with random input perturbations.

For better clarity, the proofs of some important theorems will be presented in the respective chapters, while the proofs of other results will be omitted and referred to the corresponding publication.

1.5 Notation Description

$\mathbb{R}$	the set of real numbers.
$\mathbb{R}_+$	the set of positive real numbers.
$\mathbb{R}^N$	N-dimensional real vector space.
$\mathbb{R}^{N\times M}$	$N \times M$-dimensional real matrix space.

$\mathbb{R}_+^{\mathrm{N}}$	the set of all non-negative N-dimensional vectors.
$\mathbb{N}$	the set of natural numbers.
$\mathbb{C}^N$	N-dimensional complex space.
$\mathbb{E}\{\}$	mathematic expectation.
$\mathbf{A}$	matrices.
$\mathbf{a}$	column vectors.
(P, Q)	power flow vectors.
$[\mathrm{X}]_{i,j}$	(i, j)-th entry of matrix $\mathbf{X}$.
$[\mathrm{X}]_{i:}$	i-th row of matrix $\mathbf{X}$.
$[\mathrm{X}]_{:j}$	j-th column of matrix $\mathbf{X}$.
$(\cdot)^{\mathrm{T}}$	transpose of the matrix.
$(\cdot)^{\mathrm{H}}$	Hermitian transpose of the matrix.
$\mathrm{Tr}(\cdot)$	trace of the matrix.
$\overline{(\cdot)}$	complex conjugate of the matrix.
$\Re(\cdot)$	the real part of a complex number.
$\Im(\cdot)$	the real imaginary part of a complex number.
$\|x\|$	the l_2-norm of vector x.
$\mathbf{0}$	all-zeros vectors of suitable dimensions.
$\mathbf{1}$	all-ones vectors of suitable dimensions.
I	the identity matrix of the appropriate dimension.
$[\mathrm{x}_i]_{1\le i\le N}$	a matrix with $\mathrm{x}_i^{\mathrm{H}}$ being its i-th row.
$\mathrm{dg}(\mathrm{x}_i)$	a diagonal matrix holding entries of x_i on its diagonal.
$\triangle(\mathcal{S})$	the distribution over space $\mathcal{S}$.

Chapter 2
State Estimation via Composite Optimization

Precise monitoring of the grid's operational state holds paramount significance for a multitude of control and optimization endeavors. These optimization tasks such as optimal power flow, reliability analysis, attack detection, and future network expansion planning (Abur and Exposito 2004; Zhang et al. 2019a). In a groundbreaking publication (Schweppe and Wildes 1970), the fundamental statistical framework was established for the field of PSSE. The core objective of PSSE is to ascertain the voltage magnitudes and angle information at all network buses, leveraging network parameters and measurements collected from various points within the grid. SCADA systems serve as the primary means for monitoring power grids. However, the presence of parameter uncertainty, instrument mis-calibration, and unmonitored topology alterations can lead to the generation of severely tainted SCADA measurements, often referred to as 'bad data' (Merrill and Schweppe 1971). In the context of the smart grid, there is a clear impetus for the development of efficient and robust PSSE methodologies to mitigate cybersecurity threats (Zonouz et al. 2012).

Commonly used PSSE criteria include the WLS and the LAV (Caro and Conejo 2012). Additionally, for the pursuit of enhanced robustness, alternative estimators come into play, such as the Schweppe-Huber generalized M-Estimator (Mili et al. 1996b; Zhao and Mili 2018; Zhao et al. 2018), along with the least-median and the least-trimmed squares state estimators (Mili et al. 1996b). On the other hand, LAV estimators fulfill the dual role of identifying and excluding erroneous data points while simultaneously yielding a precise state estimate (Kotiuga and Vidyasagar 1982). The LAV-based PSSE is revisited in this work from the viewpoint of composite optimization (Burke 1985; Lewis and Wright 2016). This perspective centers around the minimization of functions of the form $f(\mathbf{v}) = h(c(\mathbf{v}))$, where these functions are compositions involving a convex function h and a differentiable vector function c. While the presence of problematic leverage points may pose challenges, it is noteworthy that the proposed prox-linear algorithms can be extended to encompass robust estimation formulations. These encompass the Huber estimation, Huber

G. Wang et al., *Data-driven Optimization and Control for Autonomous Energy Systems*,
https://doi.org/10.1007/978-981-95-1782-4_2

M-estimation, and the Schweppe-Huber generalized M-estimation methods, as documented in references (Mili et al. 1996b; Gandhi and Mili 2009; Pires et al. 1999; Zhao and Mili 2018; Zhao et al. 2018).

In this chapter, we draw inspiration from the latest developments in composite optimization and introduce two effective LAV estimation algorithms. The first algorithm employs a deterministic linear proximal approach, tackling a series of convex quadratic problems (typically numbering between 5 and 10), each efficiently solvable through established toolsets or the ADMM. The second algorithm takes a stochastic approach, updating only a few elements of the intricate voltage state vector during per iteration. Remarkably, even when exclusively utilizing voltage magnitudes and active power flow measurements, this number of updated elements remains as minimal as one or two, regardless of the network's bus count. Such computational efficiency is well-suited for large-size power systems. Furthermore, by carefully mini-batching the voltage and power flow measurements, accelerated implementation of the stochastic iterations becomes possible. These proposed algorithms are subjected to numerical assessments using various benchmark power grids. The outcomes of simulation tests unequivocally demonstrate substantial reductions in computation time and a marked enhancement in robustness compared to existing alternative approaches, particularly within medium- to large-size networks.

2.1 System Modeling and Problem Formulation

An electric grid having N buses and L lines is represented as a graph $\mathcal{G} = (\mathcal{N}, \mathcal{E})$. In this graph, the set of nodes $\mathcal{N} := \{1, 2, \ldots, N\}$ corresponds to the buses, while the set of edges $\mathcal{E} := \{(n, n')\} \subseteq \mathcal{N} \times \mathcal{N}$ corresponds to the lines. The complex voltage at each bus $n \in \mathcal{N}$ is denoted in rectangular coordinates as $v_n = \Re(v_n) + j\Im(v_n)$, where all nodal voltages collectively form the vector $\mathbf{v} := [v_1 \cdots v_N]^{\mathrm{H}} \in \mathbb{C}^N$.

The voltage magnitude square $V_n := |v_n|^2 = \Re^2(v_n) + \Im^2(v_n)$ can be compactly expressed as a quadratic function of $\mathbf{v}$, namely

$$V_n = \mathbf{v}^{\mathrm{H}}\mathbf{H}_n^V\mathbf{v}, \text{ with } \mathbf{H}_n^V := \mathbf{e}_n\mathbf{e}_n^{\mathrm{T}}, \tag{2.1}$$

where $\mathbf{e}_n$ denotes the n-th canonical vector in $\mathbb{R}^N$. To express power injections as functions of $\mathbf{v}$, introduce the so-called bus admittance matrix $\mathbf{Y} = \mathbf{G} + j\mathbf{B} \in \mathbb{C}^N$ (Abur and Exposito 2004). In rectangular coordinates, the active and reactive power injections p_n and q_n at bus n are given by

$$p_n = \Re(v_n)\sum_{n'=1}^{N}[\Re(v_{n'})G_{nn'} - \Im(v_n)B_{nn'}] + \Im(v_n)\sum_{n'=1}^{N}[\Im(v_{n'})G_{nn'} + \Re(v_{n'})B_{nn'}], \tag{2.2}$$

$$q_n = \Im(v_n)\sum_{n'=1}^{N}[\Re(v_{n'})G_{nn'} - \Im(v_n)B_{nn'}] - \Re(v_n)\sum_{n'=1}^{N}[\Im(v_{n'})G_{nn'} + \Re(v_{n'})B_{nn'}], \tag{2.3}$$

which can be compactly represented as

$$p_n = \mathbf{v}^{\mathrm{H}}\mathbf{H}_n^p\mathbf{v}, \text{ with } \mathbf{H}_n^p := \frac{\mathbf{Y}^{\mathrm{H}}\mathbf{e}_n\mathbf{e}_n^{\mathrm{T}} + \mathbf{e}_n\mathbf{e}_n^{\mathrm{T}}\mathbf{Y}}{2}, \tag{2.4a}$$

$$q_n = \mathbf{v}^{\mathrm{H}}\mathbf{H}_n^q\mathbf{v}, \text{ with } \mathbf{H}_n^q := \frac{\mathbf{Y}^{\mathrm{H}}\mathbf{e}_n\mathbf{e}_n^{\mathrm{T}} - \mathbf{e}_n\mathbf{e}_n^{\mathrm{T}}\mathbf{Y}}{2j}. \tag{2.4b}$$

Recognize that the line current from bus n to n' at the 'from' end obeys $I_{nn'} = \mathbf{e}_{nn'}^{\mathrm{T}}\mathbf{i}_f = \mathbf{e}_{nn'}^{\mathrm{T}}\mathbf{Y}_f\mathbf{v}$, where $\mathbf{i}_f \in \mathbb{C}^{|\mathcal{E}|}$ collects all line currents, and $\mathbf{Y}_f \in \mathbb{C}^{|\mathcal{E}|\times N}$ relates the bus voltages to all line currents at the 'from' (sending) end. Ohm's and Kirchhoff's laws assert that the 'from-end' power flow over line (n, n') can be expressed as $\overline{S}_{nn'}^f = P_{nn'}^f - jQ_{nn'}^f = \overline{v}_n i_{nn'}^f = (\mathbf{v}^{\mathrm{H}}\mathbf{e}_n)(\mathbf{e}_{nn'}^{\mathrm{T}}\mathbf{i}_f) = \mathbf{v}^{\mathrm{H}}\mathbf{e}_n\mathbf{e}_{nn'}^{\mathrm{T}}\mathbf{Y}_f\mathbf{v}$, yielding

$$P_{nn'}^f = \mathbf{v}^{\mathrm{H}}\mathbf{H}_{nn'}^P\mathbf{v}, \text{ with } \mathbf{H}_{nn'}^P := \frac{\mathbf{Y}_f^{\mathrm{H}}\mathbf{e}_{nn'}\mathbf{e}_n^{\mathrm{T}} + \mathbf{e}_n\mathbf{e}_{nn'}^{\mathrm{T}}\mathbf{Y}_f}{2}, \tag{2.5a}$$

$$Q_{nn'}^f = \mathbf{v}^{\mathrm{H}}\mathbf{H}_{nn'}^Q\mathbf{v}, \text{ with } \mathbf{H}_{nn'}^Q := \frac{\mathbf{Y}_f^{\mathrm{H}}\mathbf{e}_{nn'}\mathbf{e}_n^{\mathrm{T}} - \mathbf{e}_n\mathbf{e}_{nn'}^{\mathrm{T}}\mathbf{Y}_f}{2j}, \tag{2.5b}$$

The active and reactive power flows, denoted as $P_{nn'}^t$ and $Q_{nn'}^t$, measured at the 'to' (receiving) ends can be symmetrically expressed as $P_{nn'}^f$ and $Q_{nn'}^f$. For the sake of conciseness, we omit their detailed presentation here.

With the line parameters in $\mathbf{Y}$ and $\mathbf{Y}_f$, all SCADA measurements, including voltage magnitudes, active and reactive power injections, and power flows, can be expressed as quadratic functions of the voltage vector $\mathbf{v} \in \mathbb{C}^N$. This substantiates why $\mathbf{v}$ is commonly referred to as the system state. If we denote $\mathcal{S}_V$, $\mathcal{S}_p$, $\mathcal{S}_q$, $\mathcal{S}_P^f$, $\mathcal{S}_Q^f$, $\mathcal{S}_P^t$, and $\mathcal{S}_Q^t$ as the smart meter locations corresponding to their respective types, we are equipped with the following set of measurements, which may occasionally be subject to noise or potential corruption: $\{\breve{V}_n\}_{n\in\mathcal{S}_V}$, $\{\breve{p}_n\}_{n\in\mathcal{S}_p}$, $\{\breve{q}_n\}_{n\in\mathcal{S}_q}$, $\{\breve{P}_{nn'}^f\}_{(n,n')\in\mathcal{S}_P^f}$, $\{\breve{Q}_{nn'}^f\}_{(n,n')\in\mathcal{S}_Q^f}$, $\{\breve{P}_{nn'}^t\}_{(n,n')\in\mathcal{S}_P^t}$, and $\{\breve{Q}_{nn'}^t\}_{(n,n')\in\mathcal{S}_Q^t}$, henceforth concatenated in the vector $\mathbf{z} \in \mathbb{R}^M$, where M denotes the total number of measurements.

In this section, the following corruption model is considered in (Duchi and Ruan 2019): If $\{\xi_i\} \subseteq \mathbb{R}$ models an arbitrary attack (or outlier) sequence, given the measurement matrices $\{\mathbf{H}_m\}_{m=1}^M$ in (2.1–2.5a), we observe for $1 \leq m \leq M$ the samples

$$z_m \approx \begin{cases} \mathbf{v}^{\mathrm{H}}\mathbf{H}_m\mathbf{v} & \text{if } m \in \mathcal{I}^{nom}, \\ \xi_m & \text{if } m \in \mathcal{I}^{out}, \end{cases} \tag{2.6}$$

where additive measurement noise can be included if $\approx$ is replaced by equality, and $\mathcal{I}^{nom}, \mathcal{I}^{out} \subseteq \{1, 2, \ldots, M\}$ collect the indices of nominal data and outliers, respectively. To elaborate further, $\mathcal{I}^{out}$ denotes the collection of meter indices that are susceptible to compromise. These indices, which belong to $\mathcal{I}^{out}$, are presumed to be selected in a stochastic manner from the set $\{1, 2, \ldots, M\}$ It's essential to note that instrument failures transpire randomly, while the sequence of attacks, denoted as $\{\xi_m\}$, may potentially exhibit a dependency on $\{\mathbf{H}_m\}$, and this interrelation could even take on an adversarial character. To delve into the specifics, we will delve into two distinct models for these attacks.

M1 Matrices $\{\mathbf{H}_m\}_{m=1}^M$ are independent of $\{\xi_m\}_{m=1}^M$.

M2 Nominal measurement matrices $\{\mathbf{H}_m\}_{m\in\mathcal{I}^{nom}}$ are independent of $\{\xi_m\}_{m\in\mathcal{I}^{out}}$.

It is crucial to emphasize that M1 necessitates complete independence between the corruption and measurements. In other words, the attacker is only permitted to corrupt ξ_m without any knowledge of $\mathbf{H}_m$. In stark contrast, M2 accommodates entirely arbitrary dependencies between ξ_m and $\mathbf{H}_m$ for instances where m belongs to the set $\mathcal{I}^{out}$. This allowance aligns with the natural case where the nature of corruption may indeed rely on the specific characteristics of the individual measurement $\mathbf{H}_m$.

Having provided detailed explanations of the system and corruption models, we can now articulate the PSSE problem as follows: Given matrices $\mathbf{Y}$ and $\mathbf{Y}_f$, as well as the set of available measurements $\mathbf{z} \in \mathbb{R}^M$, with entries as defined in (2.6) following either M1 or M2, the objective is to recover the voltage vector $\mathbf{v} \in \mathbb{C}^N$. Initially, one might consider seeking the WLS estimate, or, in the case of assuming independent Gaussian noise, the ML estimate (Schweppe and Wildes 1970). However, it is worth noting that the WLS criterion is susceptible to the influence of outliers and can produce highly inaccurate estimates, even when only a few measurements are significantly corrupted (Merrill and Schweppe 1971). As extensively documented in the fields of statistics and optimization, ℓ_1-based loss functions lead to estimators based on medians (Huber 2004) and offer a more robust handling of gross errors in the measurements $\mathbf{z}$. Motivated by this, we will explore the minimization of ℓ_1 loss in the residuals, which leads to what is known as the LAV estimate (Kotiuga and Vidyasagar 1982).

$$\min_{\mathbf{v}\in\mathbb{C}^N} f(\mathbf{v}) := \frac{1}{M}\sum_{m=1}^{M} \left|\mathbf{v}^{\mathrm{H}}\mathbf{H}_m\mathbf{v} - z_m\right|. \tag{2.7}$$

The objective function in (2.7), associated with the LAV estimator, presents distinct characteristics due to the presence of $\{\mathbf{v}^{\mathrm{H}}\mathbf{H}_m\mathbf{v}\}_{m=1}^M$ and the absolute-value operation. It is noteworthy that this objective function is nonsmooth, nonconvex, and notably lacks local convexity near the optimal solutions $\pm\mathbf{v}^*$. This is evident when considering a real-valued scalar function $f(v) = |v^*v-1|$, where $v \in \mathbb{R}$. In such cases, conducting a local analysis based on convexity and smoothness becomes impractical, making the minimization of $f(\mathbf{v})$ a challenging endeavor. Consequently, traditional optimization techniques like Gauss–Newton are unsuitable for minimizing (2.7). However, it is worth highlighting that the criterion $f(\mathbf{v})$ possesses several unique structural properties, which we will leverage in the following sections to develop efficient optimization algorithms.

Remark 2.1 For an N-bus power system, the majority of existing PSSE approaches have traditionally focused on optimizing a set of $(2N - 1)$ real variables. These variables encompass either the polar or rectangular components of the complex voltage phasors, with the exception of the angle or the imaginary part of the reference bus, which is typically set to 0. However, when employing iterative algorithms, it is generally more efficient and computationally advantageous to directly operate with the N-dimensional complex voltage vector. This preference arises from the inherent advantages offered by complex voltage phasors, including their compact quadratic representations for all SCADA quantities. Notably, these representations exhibit natural sparsity in the quadratic measurement matrices associated with the unknown complex voltage phasor vector.

2.2 Deterministic Prox-Linear Least-Absolute Value Solver

In this section, we will formulate a deterministic solver for (2.7). To initiate this process, we shall begin by re-expressing the objective function presented in (2.7) as follows

$$\min_{\mathbf{v}\in\mathbb{C}^N} f(\mathbf{v}) := h(c(\mathbf{v})), \tag{2.8}$$

the formulation comprises a convex function $h : \mathbb{R}^M \to \mathbb{R}$ and a smooth vector function $c : \mathbb{C}^N \to \mathbb{R}^M$, a structure that is well-recognized for its compatibility with efficient algorithms (Burke 1985, Lewis and Wright 2016). It is evident that this general structure encompasses (2.7) as a special instance, where we can specify $h(\mathbf{u}) = (1/M)\|\mathbf{u}\|_1$ and $c(\mathbf{v}) = \left[\mathbf{v}^{\mathrm{H}}\mathbf{H}_m\mathbf{v} - z_m\right]_{1\le m\le M}$. The compositional nature of this structure makes it amenable to the proximal linear (prox-linear) algorithm, which is a variant of the Gauss–Newton iterations (Burke 1985). To be precise, in the vicinity of a given $\mathbf{v}$, we define the local 'linearization' of f as follows

$$f_{\mathbf{v}}(\mathbf{w}) := h(c(\mathbf{v}) + \Re(\nabla^{\mathrm{H}}c(\mathbf{v})(\mathbf{w} - \mathbf{v}))), \tag{2.9}$$

where $\nabla c(\mathbf{v}) \in \mathbb{C}^{N\times M}$ denotes the Jacobian matrix of c at $\mathbf{v}$ based on Wirtinger derivatives for functions of complex-valued variables (Wang et al. 2018b). It's worth emphasizing that, in contrast to the nonconvex function $f(\mathbf{v})$, the function $f_{\mathbf{v}}(\mathbf{w})$ in (2.9) exhibits convexity concerning $\mathbf{w}$. This convexity property is pivotal and forms the foundation of the prox-linear method. To initiate the iterative process, an initial point denoted as $\mathbf{v}_0$ can be selected, often chosen as the flat-voltage profile point, typically represented by the all-one vector. With this starting point established, we construct the iteration process as follows

$$\mathbf{v}_{t+1} := \arg\min_{\mathbf{v}\in\mathbb{C}^N}\left\{f_{\mathbf{v}_t}(\mathbf{v}) + \frac{1}{2\mu_t}\|\mathbf{v}-\mathbf{v}_t\|_2^2\right\}, \tag{2.10}$$

where $\mu_t > 0$ is a stepsize that can be fixed in advance, or be determined by a line search (Lewis and Wright 2016).

Clearly, the subproblem (2.10), which must be tackled at per iteration of the prox-linear algorithm, possesses inherent convexity and can be efficiently handled using existing solvers such as CVX (Grant et al. 2009). However, it's important to recognize that these interior-point solvers may face scalability challenges, particularly when confronted with large sets of $\{\mathbf{H}_m\}$. To address this concern, we will now introduce a more efficient iterative approach by harnessing ADMM iterations (Boyd et al. 2011, Kekatos and Giannakis 2012).

When designating f as the LAV objective in accordance with (2.7), the minimization in (2.10) becomes

$$\mathbf{v}_{t+1} = \arg\min_{\mathbf{v}\in\mathbb{C}^N}\|\Re(\mathbf{A}_t(\mathbf{v}-\mathbf{v}_t)) - \mathbf{c}_t\|_1 + \frac{1}{2}\|\mathbf{v}-\mathbf{v}_t\|_2^2, \tag{2.11}$$

with coefficients given by

$$\mathbf{A}_t := \left[(2\mu_t/M)\mathbf{v}_t^{\mathrm{H}}\mathbf{H}_m\right]_{1\le m\le M}, \tag{2.12a}$$

$$\mathbf{c}_t := \left[(\mu_t/M)(z_m - \mathbf{v}_t^{\mathrm{H}}\mathbf{H}_m\mathbf{v}_t)\right]_{1\le m\le M}. \tag{2.12b}$$

In the interest of conciseness, let us introduce the shorthand notation $\mathbf{w} := \mathbf{v} - \mathbf{v}_t$, and rewrite (2.11) equivalently as a constrained optimization problem

$$\min_{\mathbf{u}\in\mathbb{C}^M,\ \mathbf{w}\in\mathbb{C}^N}\|\Re(\mathbf{u}) - \mathbf{c}_t\|_1 + \frac{1}{2}\|\mathbf{w}\|_2^2, \tag{2.13a}$$

$$\text{subject to}\quad \mathbf{A}_t\mathbf{w} = \mathbf{u}. \tag{2.13b}$$

To further facilitate the decoupling of constraints and streamline the ADMM implementation, we introduce auxiliary variables, denoted as $\tilde{\mathbf{u}}$ and $\tilde{\mathbf{w}}$, to correspond with $\mathbf{u}$ and $\mathbf{w}$, respectively. Consequently, we can reformulate (2.13a) in the following manner

$$\min_{\tilde{\mathbf{u}},\tilde{\mathbf{w}},\mathbf{u},\mathbf{w}}\|\Re(\tilde{\mathbf{u}}) - \mathbf{c}_t\|_1 + \frac{1}{2}\|\tilde{\mathbf{w}}\|_2^2, \tag{2.14a}$$

$$\text{subject to}\,\tilde{\mathbf{u}} = \mathbf{u},\ \tilde{\mathbf{w}} = \mathbf{w},\ \mathbf{A}_t\mathbf{w} = \mathbf{u}. \tag{2.14b}$$

Introducing $\lambda \in \mathbb{C}^N$ and $\nu \in \mathbb{C}^M$ as the Lagrange multipliers associated with the **w**- and **u**- consensus constraints, respectively, the augmented Lagrangian, with the final equality in (2.14b) omitted, can be represented as

$$\begin{aligned}\mathcal{L}(\tilde{\mathbf{w}}, \tilde{\mathbf{u}}, \mathbf{w}, \mathbf{u}; \lambda, \nu) : = \|\Re(\tilde{\mathbf{u}}) - \mathbf{c}_t\|_1 + \frac{1}{2}\|\tilde{\mathbf{w}}\|_2^2 + \Re(\lambda^{\mathrm{H}}(\tilde{\mathbf{w}} - \mathbf{w})) + \Re(\nu^{\mathrm{H}}(\tilde{\mathbf{u}} - \mathbf{u}))\\ + \frac{\rho}{2}\|\tilde{\mathbf{w}} - \mathbf{w}\|_2^2 + \frac{\rho}{2}\|\tilde{\mathbf{u}} - \mathbf{u}\|_2^2 \end{aligned}\tag{2.15}$$

where $\rho > 0$ denotes a predefined step size. Let $k \in \mathbb{N}$ represent the iteration index for solving (2.13a), which is equivalent to (2.10). In each ADMM iteration, the following recursions are performed

$$\tilde{\mathbf{w}}^{k+1} := \arg\min_{\tilde{\mathbf{w}}}\left\{\frac{1}{2}\|\tilde{\mathbf{w}}\|_2^2 + \frac{\rho}{2}\left\|\tilde{\mathbf{w}} - (\mathbf{w}^k - \lambda^k)\right\|_2^2\right\}, \tag{2.16a}$$

$$\tilde{\mathbf{u}}^{k+1} := \arg\min_{\tilde{\mathbf{u}}}\left\{\frac{1}{2}\|\Re(\tilde{\mathbf{u}}) - \mathbf{c}_t\|_1 + \frac{\rho}{2}\left\|\tilde{\mathbf{u}} - (\mathbf{u}^k - \nu^k)\right\|_2^2\right\}, \tag{2.16b}$$

$$\left\{\mathbf{w}^{k+1}, \mathbf{u}^{k+1}\right\} := \arg\min_{\mathbf{w},\,\mathbf{u}}\left\|\mathbf{w} - (\tilde{\mathbf{w}}^{k+1} + \lambda^k)\right\|_2^2 + \left\|\mathbf{u} - (\tilde{\mathbf{u}}^{k+1} + \nu^k)\right\|_2^2, \tag{2.16c}$$

$$\text{subject to}\,\mathbf{A}_t\mathbf{w} = \mathbf{u}, \tag{2.16d}$$

$$\begin{bmatrix}\lambda^{k+1}\\ \nu^{k+1}\end{bmatrix} = \begin{bmatrix}\lambda^k + (\tilde{\mathbf{w}}^{k+1} - \mathbf{w}^{k+1})\\ \nu^k + (\tilde{\mathbf{u}}^{k+1} - \mathbf{u}^{k+1})\end{bmatrix}, \tag{2.16e}$$

where all the dual variables have been scaled by the factor $\rho > 0$ (Boyd et al. 2011).

Proposition 2.1 *The solutions of* (2.16a) *and* (2.16b) *are respectively.*

$$\tilde{\mathbf{w}}^{k+1} := \frac{\rho}{1+\rho}(\mathbf{w}^k - \lambda^k), \tag{2.17a}$$

$$\tilde{\mathbf{u}}^{k+1} := \mathbf{c}_t + \mathcal{S}_{1/2\rho}(\Re(\mathbf{u}^k - \nu^k) - \mathbf{c}_t) + i\Im(\mathbf{u}^k - \nu^k), \tag{2.17b}$$

where the shrinkage operator $\mathcal{S}_\tau(\mathbf{x}) : \mathbb{R}^N \times \mathbb{R}_+ \to \mathbb{R}^N$ is $\mathcal{S}_\tau(\mathbf{x}) := \text{sign}(\mathbf{x}) \odot \max(|\mathbf{x}| - \tau\mathbf{1}, 0)$, with $\odot$ and $|\cdot|$ denoting the entrywise multiplication and absolute operators, respectively, $\text{sign}(x) := \begin{cases} x/|x|, & x \neq 0,\\ 0, & x = 0,\end{cases}$ provides an entrywise definition of operator sign(**x**).

Proof The solution to (2.16a) can be readily obtained from (2.17a), and as such, its proof is omitted here. When dealing with arbitrary vectors $\mathbf{c} \in \mathbb{R}^N$ and $\mathbf{d} \in \mathbb{C}^N$, the task of solving (2.16b) is analogous to solving the following equation.

$$\mathbf{u}^* := \arg\min_{\mathbf{u}\in\mathbb{C}^N} \lambda\|\Re(\mathbf{u}) - \mathbf{c}\|_1 + \frac{1}{2}\|\mathbf{u} - \mathbf{d}\|_2^2. \tag{2.18}$$

Upon defining $\mathbf{x} := \mathbf{u} - \mathbf{c}$, problem (2.18) becomes

$$\min_{\mathbf{x}\in\mathbb{C}^N} \ \lambda\|\Re(\mathbf{x})\|_1 + \frac{1}{2}\|\mathbf{x} - (\mathbf{d} - \mathbf{c})\|_2^2,$$

or equivalently,

$$\min_{\mathbf{x}} \lambda\|\Re(\mathbf{x})\|_1 + \frac{1}{2}\|\Re(\mathbf{x}) - \Re(\mathbf{d} - \mathbf{c}))\|_2^2 + \frac{1}{2}\|\Im(\mathbf{x}) - \Im(\mathbf{d} - \mathbf{c}))\|_2^2. \tag{2.19}$$

Problem (2.19) can be decomposed into two distinct subproblems, each corresponding to the optimization of the real and imaginary components of $\mathbf{x} = \Re(\mathbf{x}) + j\Im(\mathbf{x}) := \mathbf{x}_r + j\mathbf{x}_i$. This decomposition is expressed as

$$\min_{\mathbf{x}_r\in\mathbb{R}^N} \ \lambda\|\mathbf{x}_r\|_1 + \frac{1}{2}\|\mathbf{x}_r - \Re(\mathbf{d} - \mathbf{c})\|_2^2, \tag{2.20}$$

and

$$\min_{\mathbf{x}_i\in\mathbb{R}^N} \frac{1}{2}\|\mathbf{x}_i - \Im(\mathbf{d} - \mathbf{c})\|_2^2, \tag{2.21}$$

The optimal solutions of the convex programs in (2.20) and (2.21) can be found as, see e.g. (Boyd et al. 2011)

$$\begin{aligned}\mathbf{x}_r^* &:= \mathcal{S}_\lambda(\Re(\mathbf{d} - \mathbf{c})) = \mathcal{S}_\lambda(\Re(\mathbf{d}) - \mathbf{c}),\\ \mathbf{x}_i^* &:= \Im(\mathbf{d} - \mathbf{c}) = \Im(\mathbf{d}),\end{aligned}$$

thus yielding the optimal solution of (2.19) as

$$\mathbf{x}^* := \mathbf{x}_r^* + j\mathbf{x}_i^* = \mathcal{S}_\lambda(\Re(\mathbf{d}) - \mathbf{c}) + j\Im(\mathbf{d}).$$

Recalling $\mathbf{u} = \mathbf{x} + \mathbf{c}$, the optimal solution of (2.18) is

$$\mathbf{u}^* = \mathbf{x}^* + \mathbf{c} = \mathbf{c} + \mathcal{S}_\lambda(\Re(\mathbf{d}) - \mathbf{c}) + j\Im(\mathbf{d}), \tag{2.22}$$

which completes the proof.

The constrained minimization problem presented in (2.16d) essentially involves projecting the pair $(\tilde{\mathbf{w}}^{k+1} + \lambda^k, \tilde{\mathbf{u}}^{k+1} + \nu^k)$ onto the convex set defined by the linear equality constraint, namely $\{(\mathbf{w}, \mathbf{u}) : \mathbf{A}_t\mathbf{w} = \mathbf{u}\}$. The solution to this problem can be derived in a straightforward closed-form manner, as detailed below.

Proposition 2.2 *Given* $\mathbf{b} \in \mathbb{C}^N$ *and* $\mathbf{d} \in \mathbb{C}^M$, *the solution of*

$$\min_{\mathbf{w}\in\mathbb{C}^N,\,\mathbf{u}\in\mathbb{C}^M} \frac{1}{2}\|\mathbf{w}-\mathbf{b}\|_2^2 + \frac{1}{2}\|\mathbf{u}-\mathbf{d}\|_2^2,$$
$$\text{subject to } \mathbf{A}\mathbf{w}=\mathbf{u}$$

is given as

$$\mathbf{w}^* := \left(\mathbf{I}+\mathbf{A}^{\mathrm{H}}\mathbf{A}\right)^{-1}\left(\mathbf{b}+\mathbf{A}^{\mathrm{H}}\mathbf{d}\right), \tag{2.23a}$$

$$\mathbf{u}^* := \mathbf{A}\mathbf{w}^*. \tag{2.23b}$$

Proof Letting χ denote the dual variable associated with the constraint $\mathbf{u}=\mathbf{A}\mathbf{w}$, the Karush–Kuhn–Tucker optimality conditions are given by (Boyd et al. 2011).

$$\begin{aligned}
&\mathbf{w}^* - \mathbf{b} + \mathbf{A}^{\mathrm{H}}\chi^* = \mathbf{0},\\
&\mathbf{u}^* - \mathbf{d} - \chi^* = \mathbf{0},\\
&\mathbf{A}\mathbf{w}^* - \mathbf{u}^* = \mathbf{0},
\end{aligned}$$

or in the following compact form

$$\begin{bmatrix} \mathbf{I}_N & \mathbf{0} & \mathbf{A}^{\mathrm{H}} \\ \mathbf{0} & \mathbf{I}_M & -\mathbf{I}_M \\ \mathbf{A} & -\mathbf{I}_M & \mathbf{0} \end{bmatrix} \begin{bmatrix} \mathbf{w}^* \\ \mathbf{u}^* \\ \chi^* \end{bmatrix} = \begin{bmatrix} \mathbf{b} \\ \mathbf{d} \\ \mathbf{0} \end{bmatrix}.$$

Eliminating the dual variable via $\chi^* = \mathbf{u}^* - \mathbf{d}$ from the Karush–Kuhn–Tucker system, yields

$$\begin{bmatrix} \mathbf{I}_N & \mathbf{A}^{\mathrm{H}} \\ \mathbf{A} & -\mathbf{I}_M \end{bmatrix} \begin{bmatrix} \mathbf{w}^* \\ \mathbf{u}^* \end{bmatrix} = \begin{bmatrix} \mathbf{b}+\mathbf{A}^{\mathrm{H}}\mathbf{d} \\ \mathbf{0} \end{bmatrix}. \tag{2.24}$$

By further eliminating $\mathbf{d}^*$ and solving for $\mathbf{b}^*$, the solution to (2.24) and also to the minimization in (2.23a–2.23b) can be found in two steps as

$$\begin{aligned}
&\mathbf{w}^* := \left(\mathbf{I}+\mathbf{A}^{\mathrm{H}}\mathbf{A}\right)^{-1}\left(\mathbf{b}+\mathbf{A}^{\mathrm{H}}\mathbf{d}\right),\\
&\mathbf{u}^* := \mathbf{A}\mathbf{w}^*,
\end{aligned}$$

which completes the proof of the claim.

Using Proposition 2.2, the minimizer of (2.16d) is found as

$$\mathbf{w}^{k+1} := (\mathbf{I}+\mathbf{A}_t^{\mathrm{H}}\mathbf{A}_t)^{-1}\left[\tilde{\mathbf{w}}^{k+1} + \lambda^k + \mathbf{A}^{\mathrm{H}}(\tilde{\mathbf{u}}^{k+1} + \nu^k)\right], \tag{2.25a}$$

$$\mathbf{u}^{k+1} := \mathbf{A}\mathbf{w}^{k+1}. \tag{2.25b}$$

The four updates specified in (2.16a) are computationally straightforward, with the exception of the matrix inversion task in (2.25a). However, it's important to note that this inversion can be computed once and cached for subsequent iterations after its initial calculation. Furthermore, the initialization of the ADMM variables $\tilde{\mathbf{u}}^0$, λ^0, and ν^0 to zero is a viable starting point. Lastly, the solution to (2.11) can be readily obtained as follows

$$\mathbf{v}_{t+1} := \mathbf{v}_t + \mathbf{w}^*, \tag{2.26}$$

where $\mathbf{w}^*$ is the converged $\mathbf{w}$-iterate of the ADMM iterations in (2.17a), (2.25a), and (2.16e).

The innovative deterministic LAV solver, rooted in ADMM principles, is concisely presented in Algorithm 2.1. Notably, the inner loop encompassing Steps 3 to 8 offers the flexibility to be substituted with readily available solvers for tackling (2.11) and deriving $\mathbf{v}_{t+1}$.

In terms of performance considerations, when h exhibits L-Lipschitz continuity and ∇c demonstrates β-Lipschitz continuity, adhering to a constant stepsize $\mu \leq 1/(L\beta)$ within (2.10) comes with a guarantee, as detailed in (Davis and Drusvyatskiy 2019):

(i) The proposed solver outlined in Algorithm 2.1 employs a descent method.
(ii) The iterative sequence $\{\mathbf{v}_t\}$ converges reliably to a stationary point of the LAV objective delineated in (2.7).

The computational load of the ADMM-based deterministic solver primarily hinges on the projection step within (2.25a), resulting in a per-iteration computational complexity of approximately $\mathcal{O}(MN^2)$. This complexity can be afforded in small- or medium-size PSSE tasks, but may not be efficient enough for nowadays increasingly interconnected power networks. Consequently, we will explore a more efficient stochastic alternative in the following section.

Algorithm 2.1 Deterministic LAV Solver Using ADMM

1: **Input** data $\{(z_m, \mathbf{H}_m)\}_{m=1}^M$, stepsizes $\mu, \rho > 0$, and initialization $\mathbf{v}_0 = \mathbf{1}$
2: **For** $t = 0, 1, \ldots,$ **do**
3: **Initialize** $\mathbf{w}^0$, $\mathbf{u}^0$, λ^0, and ν^0 to zero
4: **Evaluate** $\mathbf{A}_t$ and $\mathbf{c}_t$ as in (2.12a), (2.12b)
5: **For** $k = 0, 1, \ldots,$ **do**
6: **Update** $(\tilde{\mathbf{w}}^{k+1}, \tilde{\mathbf{u}}^{k+1})$, $(\mathbf{w}^{k+1}, \mathbf{u}^{k+1})$, and (μ^{k+1}, ν^{k+1}) using (2.17a), (2.17b), (2.25a), (2.25b), and (2.16e), respectively
7: **End for**
8: **Update** $\mathbf{v}_{t+1}$ via (2.26)
9: **End for**

2.3 Stochastic Prox-Linear Least-Absolute Value Solver

Determining the minimizer of (2.11) precisely per iteration of the deterministic LAV solver may pose a computationally expensive challenge, potentially reaching intractability as the network scales up significantly in size. Given the extensive utility of LAV estimation and the ever-expanding interconnection of microgrids, the pursuit of scalable, online, and stochastic approaches has garnered considerable attention. In this section, we adapt a stochastic linear proximal algorithm introduced in (Duchi and Ruan 2018) to our PSSE task, harnessing its potential to efficiently tackle the LAV estimation problem at scale. Stochastic approaches offer several advantages over their deterministic counterparts, often featuring straightforward closed-form updates and faster convergence rates that yield an (approximately) optimal solutions. Leveraging the sparsity inherent in the measurement matrices and judiciously grouping measurements into smaller mini-batches can significantly expedite the implementation of the stochastic solver.

2.3.1 *Stochastic LAV Solver*

Rather than addressing the quadratic subproblems delineated in (2.11), per iteration of the stochastic LAV solver opts for a different approach. It stochastically selects a datum m_t from the total set of measurements, denoted as $1, 2, \ldots, M$, with uniform randomness. Subsequently, it employs the functions (h_{m_t}, c_{m_t}) associated with the sampled datum to effectuate local linearization, not only in (2.9) but also within the context of (2.11), yielding

$$\mathbf{v}_{t+1} := \arg\min_{\mathbf{v}\in\mathbb{C}^N} \left\{ \left|\Re(\mathbf{a}_{m_t}^{\mathrm{H}}(\mathbf{v}-\mathbf{v}_t)) - c_{m_t}\right| + \frac{1}{2\mu_t}\|\mathbf{v}-\mathbf{v}_t\|_2^2 \right\}, \tag{2.27}$$

where the coefficients are given by

$$\mathbf{a}_{m_t} := 2\mathbf{H}_{m_t}\mathbf{v}_t, \tag{2.28a}$$

$$c_{m_t} := z_{m_t} - \mathbf{v}_t^{\mathrm{H}}\mathbf{H}_{m_t}\mathbf{v}_t. \tag{2.28b}$$

In contrast to the iterative approach employed to solve (2.11) through ADMM iterations, the minimization of (2.27) offers an elegant closed-form solution, which is elaborated upon in the subsequent result.

Proposition 2.3 *Given* $\mathbf{a} \in \mathbb{C}^N$ *and* $c \in \mathbb{R}$, *the solution of*

$$\min_{\mathbf{w}\in\mathbb{C}^N} \left|\Re(\mathbf{a}^{\mathrm{H}}\mathbf{w}) - c\right| + \frac{1}{2\tau}\|\mathbf{w}\|_2^2, \tag{2.29}$$

is given by $\mathbf{w} := \text{proj}_{\tau}(c/\|\mathbf{a}\|_2^2) \cdot \mathbf{a}$, *where the projection operator* $\text{proj}_{\tau}(x) : \mathbb{R} \times \mathbb{R}_+ \to \mathbb{R}$ *returns the real number in interval* $[-\tau, \tau]$ *closest to any given* $x \in \mathbb{R}$.

Proof The optimality condition for (2.29) is.

$$\mathbf{0} \in \partial\left(\left|\Re(\mathbf{a}^{\mathrm{H}}\mathbf{w}) - c\right|\right) + \frac{1}{\tau}\mathbf{w} \Leftrightarrow \mathbf{0} \in \partial\left|\Re(\mathbf{a}^{\mathrm{H}}\mathbf{w}) - c\right| \cdot \mathbf{a} + \frac{1}{\tau}\mathbf{w},$$

or equivalently,

$$\mathbf{0} \in \partial\left|\Re(\mathbf{a}^{\mathrm{H}}\mathbf{w}) - c\right| \cdot (\tau\mathbf{a}) + \mathbf{w},$$

where ∂ denotes the subdifferential. Let us first examine the case where $\Re(\mathbf{a}^{\mathrm{H}}\mathbf{w}) - c \neq 0$. We thus have $\partial|\Re(\mathbf{a}^{\mathrm{H}}\mathbf{w}) - c| = \text{sign}(\Re(\mathbf{a}^{\mathrm{H}}\mathbf{w}) - c)$, which yields the optimum

$$\mathbf{w}^* = -\tau \text{sign}(\Re(\mathbf{a}^{\mathrm{H}}\mathbf{w}) - c) \cdot \mathbf{a}.$$

Note that if $c/\|\mathbf{a}\|_2^2 \geq \tau$, or $\Re(\mathbf{a}^{\mathrm{H}}\mathbf{w}^*) - c = -\tau\|\mathbf{a}\|_2^2 \text{sign}(\mathbf{a}^{\mathrm{H}}\mathbf{w}^* - c) - c < 0$, then $\mathbf{w}^* = \tau\mathbf{a}$. Equivalently, if $c/\|\mathbf{a}\|_2^2 \leq -\tau$, or $\Re(\mathbf{a}^{\mathrm{H}}\mathbf{w}^*) - c = -\tau\|\mathbf{a}\|_2^2 \text{sign}(\mathbf{a}^{\mathrm{H}}\mathbf{w}^* - c) - c > 0$\$, then $\mathbf{w}^* = -\tau\mathbf{a}$.

If $\Re(\mathbf{a}^{\mathrm{H}}\mathbf{w}) - c = 0$, the subdifferential of the absolute-value operator belongs to the interval $[-1, 1]$; hence, the optimality condition becomes

$$\mathbf{0} \in -[-1,\ 1] \cdot (\tau\mathbf{a}) + \mathbf{w} \Leftrightarrow \mathbf{w} \in [-\tau,\ \tau]\mathbf{a}.$$

Upon letting $\text{proj}_{\tau}(x)$ denote the projection of a real number x onto the interval $[-\tau,\ \tau]$, one can combine the aforementioned three cases, and express compactly the optimum as follows

$$\mathbf{w}^* := \text{proj}_{\tau}(c/\|\mathbf{a}\|_2^2) \cdot \mathbf{a},$$

which concludes the proof of the proposition.

Based on Proposition 2.3, the solution of (2.27) is given by

$$\mathbf{v}_{t+1} := \mathbf{v}_t + \text{proj}_{\mu_t}(c_{m_t}/\left\|\mathbf{a}_{m_t}\right\|_2^2) \cdot \mathbf{a}_{m_t}. \tag{2.30}$$

In a nutshell, measurements with relatively small absolute residuals, denoted as $|c_{m_t}| \leq \left\|\mathbf{a}_{m_t}\right\|_2^2$, are categorized as 'nominal'. Consequently, the update to $\mathbf{v}_t$ involves a step size of $c_{m_t}/\left\|\mathbf{a}_{m_t}\right\|_2^2$ in alignment with the current direction of $\mathbf{a}_{m_t}$. Conversely, measurements exhibiting larger absolute residuals are likely to be outliers. In these cases, the update to $\mathbf{v}_t$ along the direction of $\mathbf{a}_{m_t}$ is constrained to a step of τ_t instead of the full $c_{m_t}/\left\|\mathbf{a}_{m_t}\right\|_2^2$.

Algorithm 2.2 Stochastic LAV Solver

1: **Input** data $\{(z_m, \mathbf{H}_m)\}_{m=1}^M$, stepsizes $\{\mu_t > 0\}$, and initialization $\mathbf{v}_0 = \mathbf{1}$
2: **For** $t = 0, 1, \ldots,$ **do**
3: **Draw** m_t uniformly at random from $\{1, 2, \ldots, M\}$, or, by cycling through $\{1, 2, \ldots, M\}$
4: **Evaluate** $\mathbf{a}_{m_t}$ and c_{m_t} as in (2.28a), (2.28b)
5: **Update** $\mathbf{v}_{t+1}$ using (2.30)
6: **End for**

The provided stochastic prox-linear LAV solver is detailed in Algorithm 2.2. To ensure convergence, it is essential to employ a diminishing stepsize sequence denoted as $\{\mu_t\}$. More precisely, we adopt stepsizes that exhibit square summability without necessarily being absolutely summable, which can be defined as follows

$$\sum_{t=0}^{\infty} \mu_t = \infty, \quad \text{and} \quad \sum_{t=0}^{\infty} \mu_t^2 < \infty. \tag{2.31}$$

As an example, a suitable choice for μ_t could be $\mu_t = \alpha t^{-\beta}$ with carefully chosen constants $\alpha > 0$ and $\beta \in (0.5, 1]$. With this selection, the sequence $\{\mathbf{v}_t\}$ converges almost surely to a stationary point of the LAV objective in (2.7) (Duchi and Ruan 2018).

In terms of computational complexity, it can be confirmed that each $\mathbf{H}_m$ matrix corresponding to a squared voltage magnitude or (re)active power flow measurement [cf. (2.1), (2.5a) and (2.5b)] contains exactly one or three nonzero entries, respectively. Consequently, if the available measurements include only these two types of measurements, computing $(\mathbf{a}_{m_t}, c_{m_t})$ and updating $\mathbf{v}_t$ per stochastic iteration require only a small number (less than or equal to 10) of scalar multiplications and additions, resulting in a per-iteration complexity of $\mathcal{O}(1)$. Importantly, this complexity is independent of the network size N. This computational complexity scales favorably even for very large interconnected power networks. It's also noteworthy that, depending on whether a voltage or power flow measurement is processed at per iteration, only one or two entries of $\mathbf{v}_t$ are updated. In contrast, even when power injections are also measured, the number of scalar operations per iteration increases to the order of the number of neighboring buses. However, this remains significantly smaller than the total network size N in the majority of real-world networks.

Remark 2.2 PMU measurements can be easily accounted for in (2.7). The developed prox-linear LAV schemes apply without any algorithmic modification.

2.3.2 *Accelerated Implementation Using Mini-batches*

While the stochastic solver presented in Algorithm 2.2 provides straightforward closed-form updates, achieving convergence in high-dimensional power networks may demand a considerable number of iterations. In recent times, stochastic approaches relying on mini-batches of measurements have gained popularity in large-scale machine learning tasks due to their potential to accelerate stochastic algorithms. However, the conventional approach of forming mini-batches through random grouping of measurements would yield a sequence of quadratic programs akin to (2.11) encountered in the deterministic LAV solver. These, however, lack closed-form solutions due to the presence of the ℓ_1 term. The innovation here lies in harnessing the inherent sparsity of $\mathbf{H}_m$ matrices to craft mini-batches in a manner that permits closed-form solutions for the resulting quadratic programs.

Suppose that active and reactive power flows over all lines and the square voltage magnitude at all buses are measured. Given that $\mathbf{H}_n^V$ in (2.1) contains exactly one nonzero entry at the (n, n)-th position, the corresponding updating vector $\mathbf{a}_n$ in (2.28a) possesses (at most) one nonzero entry at the n-th position. Consequently, when updating $\mathbf{v}_t$ using (2.30), it affects only the n-th position. This implies that refining $\mathbf{v}$ through the N voltage measurements sequentially over N stochastic iterations is essentially equivalent to updating $\mathbf{v}$ using all N measurements simultaneously within a single iteration.

In the context of power flows, every $\mathbf{H}_m$ exhibits three nonzero entries spanning two rows. To illustrate, consider $\mathbf{H}_{nn'}^P$, which features nonzero entries identified by the indices (n, n'), (n, n), and (n', n); a similar pattern is found in $\mathbf{H}_{nn'}^Q$. When dealing with the active or reactive power flow measurement over line (n, n'), the update pertains to both the n-th and n'-th entries of $\mathbf{v}_t$. Consequently, as long as each mini-batch of measurements does not overlap in terms of index positions with the others, processing a series of such mini-batch measurements one after another effectively mirrors the process of handling all measurements simultaneously in a single iteration.

For illustration, consider the IEEE 14-bus test system depicted in Fig. 2.1 (Christie 2000). Consider a total of 54 measurements, which include 14 square voltage magnitudes, as well as 20 'from-end' active and reactive power flows each. All voltage magnitudes can be grouped as a single mini-batch, or into several mini-batches by any means. One way of mini-batching each type of power flow measurements is suggested in Table 2.1, where 20 active (reactive) power flows yield 5 mini-batches of equal size. It can be easily verified that any two measurements within a group (mini-batch) are measured over two lines of non-overlapping indices.

Consider dividing all the measurements into B distinct mini-batches, represented as $\{\mathcal{B}_b\}_{b=1}^B \subseteq \{1, 2, \ldots, M\}$. In per iteration t, a mini-batch $\mathcal{B}_{b_t}$ is randomly selected from the set $\{\mathcal{B}_b\}_{b=1}^B$ with a uniform distribution. By employing this accelerated approach, which leverages mini-batching, the state estimate is updated as follows [cf. (2.30)]

$$\mathbf{v}_{t+1} := \mathbf{v}_t + \sum_{m \in \mathcal{B}_{b_t}} \text{proj}_{\mu_t}(c_m / \|\mathbf{a}_m\|_2^2) \cdot \mathbf{a}_m, \tag{2.32}$$

which is in sharp contrast to that of using ADMM iterations to deal with the quadratic subproblems (2.10) in the deterministic LAV solver.

Fig. 2.1 The IEEE 14-bus benchmark configuration

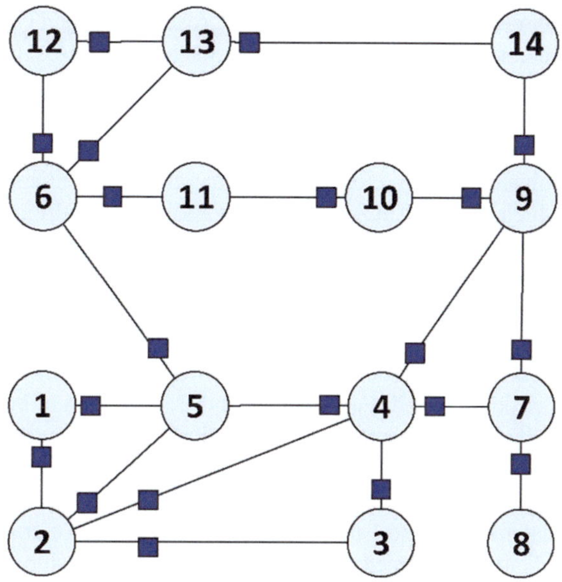

Table 2.1 Mini-batching power flow measurements

Groups	Power flow measurements (Bus index)
1	(1, 2), (3, 4), (5, 6), (7, 9)
2	(1, 5), (2, 4), (6, 11), (12, 13)
3	(2, 3), (4, 5), (6, 12), (9, 14)
4	(2, 5), (4, 7), (6, 13), (10, 11)
5	(4, 9), (7, 8), (9, 10), (13, 14)

2.4 Numerical Tests

In this section, we conducted numerical tests to evaluate the performance of the proposed linear proximal LAV solvers. We utilized three benchmark power networks: the IEEE 14-bus system, the IEEE 118-bus system, and the PEGASE 9,241-bus system. These networks were simulated using the MATLAB-based toolbox MATPOWER (Christie 2000; Zimmerman et al. 2010).

The linear programming (LP) and iteratively reweighted least-squares (IRLS) based LAV estimators (Kotiuga and Vidyasagar 1982; Abur and Celik 1991; Jabr and Pal 2004), as well as the Gauss–Newton iterations commonly used for the WLS-based PSSE (Abur and Exposito 2004), were employed as baselines for our evaluation. It is noteworthy that the LP-based implementation can be viewed as a specialized case of the deterministic prox-linear algorithm with a constant step size of ∞.

For clarification, the LP-based approach (Abur and Celik 1991) deals with the minimization problem in (2.10) but omits the augmented majorization term $\frac{1}{2\mu_t}\|\mathbf{v}-\mathbf{v}_t\|_2^2$, effectively setting $\mu_t = \infty$ per iteration.

To provide further details, the linear program was precisely formulated over $2N-1$ real variables, which encompassed both the real and imaginary components of the voltage phasor vector, with the exception of the imaginary part of the reference bus, which was fixed at zero. In per iteration, this resulting linear program was efficiently solved by employing the convex optimization package CVX (Grant et al. 2009), along with its embedded interior-point solver (SeDuMi Sturm 1999). Given that there is no parameter in the LP-based LAV estimator, although the time performance may vary if different toolboxes are used for solving the resultant linear programs, its convergence behavior in terms of the number of iterations is independent of the toolbox used. Furthermore, the Gauss–Newton iterations were implemented by calling for the embedded SE function 'doSE.m' in MATPOWER.

Regarding the initialization process, when all squared voltage magnitudes are subject to measurement, the initial point is conventionally established as the voltage magnitude vector, unless specific instructions dictate otherwise. Each simulated scheme is designed to conclude either upon reaching a maximum threshold of 100 iterations or when the normalized distance between consecutive estimates falls below a diminutive value of 10^{-10}, expressed as $\|\mathbf{v}_t - \mathbf{v}_{t-1}\|_2/\sqrt{N} \leq 10^{-10}$. To address phase ambiguity, a uniform phase assignment of 0 is imposed at the reference bus across all tests. For the sake of numerical stability and the mitigation of certain leverage measurements as discussed in reference (Göl and Abur 2014), the implemented solvers employ normalized data, specifically $\left\{\left(\frac{z_m}{\|\mathbf{H}_m\|_2}, \frac{\mathbf{H}_m}{\|\mathbf{H}_m\|_2}\right)\right\}_{m=1}^{M}$ While this work primarily centers on the development of fast and scalable implementations of LAV estimators, it is worth noting that analogous composite optimization algorithms can find applicability in other domains sharing a compositional structure similar to that of LAV (2.7). Notable examples include variants of the (robustified) Schweppe-Huber generalized M-estimator, as explored in (Mili et al. 1996b).

2.4.1 Noiseless Case

In the first experiment, we simulated noiseless data to evaluate the convergence and runtime performance of the novel algorithms compared to the WLS-based Gauss–Newton iterations, as well as the LP- and IRLS-based LAV solvers. This assessment took place on the IEEE 14-bus test system, with the default voltage profile. The utilized measurements included all 'sending-end' active and reactive power flows, along with all squared voltage magnitudes, sourced from MATPOWER (Zimmerman et al. 2010). For the ADMM-based deterministic prox-linear solver presented in Algorithm 2.1, we employed a stepsize of $\mu = 200$, and each quadratic subproblem was solved with a maximum of 150 ADMM iterations using a stepsize of $\rho = 100$.

It's noteworthy that the deterministic prox-linear solver can also be implemented using conventional convex programming techniques, involving the direct solution of subproblem (2.10). While this approach typically achieves convergence in just a few iterations (usually fewer than 10), it comes at the cost of higher computational complexity. In the stochastic algorithm presented in Algorithm 2.2, a diminishing stepsize of $1/t^{0.8}$ was employed. For the accelerated scheme described in (2.32), a stepsize of 0.8 was used, with a total of 1 mini-batches: one for all voltage magnitudes and five of equal size for (sending-end) active and reactive power flows, each grouped as outlined in Table 2.1. To assess performance, the normalized root mean-square error (RMSE), denoted as $\|\mathbf{v}_t - \mathbf{v}\|_2/\|\mathbf{v}\|_2$, was calculated at each Gauss–Newton iteration, per linear program, and every M stochastic iterations for both the stochastic and accelerated schemes. Here, $\mathbf{v}$ represents the true voltage profile, while $\mathbf{v}_t$ is the estimate obtained at the t-th iteration.

Figure 2.2 presents a comparative analysis of the normalized RMSE, involving the LP- and IRLS-based deterministic, stochastic, and accelerated LAV solvers, alongside the WLS-based Gauss–Newton iterations. The corresponding runtime and number of iterations required to meet the stopping criterion are detailed in Table 2.2. Notably, the deterministic scheme exhibits remarkable speed, accomplishing convergence to a precision of machine accuracy (10^{-16}) within just 8 iterations. The IRLS method also demonstrates agility, but akin to the WLS-based Gauss–Newton iterations, it necessitates matrix inversion per iteration, which could render it less attractive for application in extensive power systems. While the time required to solve each LP may vary depending on the software package used, the convergence behavior of the LP-based approach remains consistent in terms of the number of iterations. It's clear that solving a linear program with $2M$ constraints and $2N-1$ real variables is computationally more demanding and time-consuming compared to performing M (accelerated) stochastic LAV iterations. This observation highlights the rapid convergence rate achieved by the proposed LAV solvers, providing justification for their effectiveness.

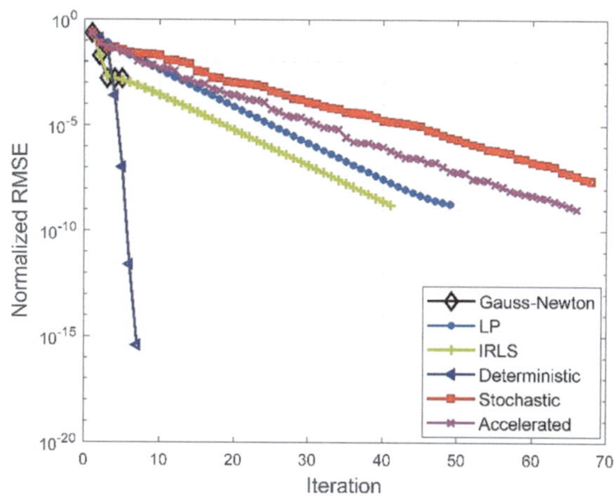

Fig. 2.2 Convergence performance for the IEEE 14-bus system

Table 2.2 Comparisons of different state estimators

Algorithm	Number of iterations	Computation time (s)
Linear program Kotiuga and Vidyasagar (1982)	49	15.167
IRLS Jabr and Pal (2004)	42	0.074
Gauss–Newton Abur and Exposito (2004)	5	0.071
Deterministic LAV	6	0.062
Stochastic LAV	68	0.110
Accelerated	66	0.103

The Gauss–Newton method concluded its computation after just six iterations, albeit at a sub-optimal point characterized by a normalized RMSE of approximately 10^{-3}. The accelerated implementation, on the other hand, exhibits performance on par with the stochastic LAV solver, delivering a precise solution with an RMSE of 4.28×10^{-8}. Remarkably, this outcome is achieved within a time frame that aligns with the duration of the Gauss–Newton iterations. It is worth emphasizing that the proposed LAV solvers significantly outpace the LP-based implementation in terms of computational speed, underscoring their efficiency and suitability for practical applications.

2.4.2 *Presence of Outliers*

The second experiment was set to assess the robustness of the novel determin istic and stochastic solvers to measurements with outliers using the IEEE 118-bus benchmark network (Christie 2000), while the IRLS-based LAV implementation and the WLS-based Gauss–Newton iterations were simulated as baselines. For this assessment, the actual voltage magnitude (in per unit) and phase angle (in radians) of each bus were generated with uniform distributions over the intervals $[0.9, 1.1]$ and $[-0.1\pi, 0.1\pi]$, respectively. To investigate the performance of PSSE concerning measurement size, an additional category of measurements was introduced systematically. Initially, all seven types of SCADA measurements were listed as $\{|V_n|^2, P^f_{nk}, Q^f_{nk}, P_n, Q_n, P^t_{nk}, Q^t_{nk}\}$. In Fig. 2.3, each value on the x-axis represents the number of ordered measurement types used in the experiment. The corresponding normalized RMSEs, which were averaged over 100 independent realizations, are depicted. For example, an x-axis value of 5 indicates that the first 5 types of data (i.e., $|V_n|^2$, P^f_{nk}, Q^f_{nk}, P_n, Q_n over all buses and lines) were included in the measurements.

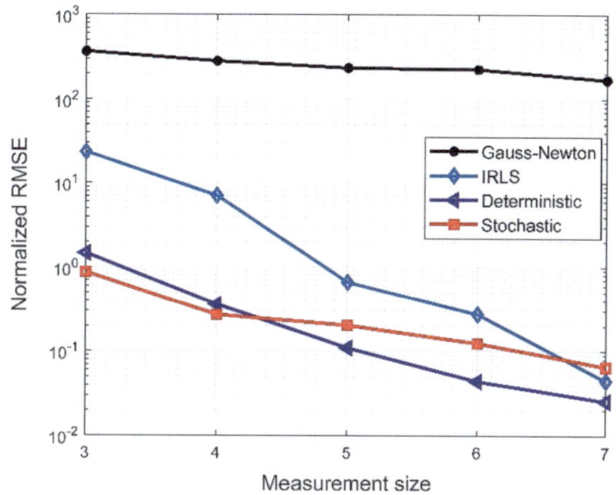

Fig. 2.3 Robustness to outliers for the IEEE 118-bus system

Additive noise was introduced independently in the measurements, following Gaussian distributions characterized by zero mean and standard deviations of 0.004 per unit for voltage magnitude, 0.008per unit for line flow, and 0.01 per unit for power injection measurements (Abur and Celik 1991). To assess robustness, ten percent of the measured data were intentionally corrupted according to the M1 model. These corruptions were randomly introduced into either line flows or bus injections. The outlying data points, denoted as $\{\xi_m\}$, were independently drawn from a Laplacian distribution featuring a zero mean and a standard deviation of 30. The deterministic scheme solved subproblems (2.10) with a fixed $\mu = 100$. This was accomplished within a maximum of 200 ADMM iterations, utilizing a stepsize of $\rho = 100$. In contrast, the stochastic approach employed a diminishing stepsize $\mu_t = 10/t^{0.9}$. The results presented in Fig. 2.3 clearly demonstrate the resilience of our prox-linear LAV schemes to measurements corrupted under model M1. These schemes exhibited improved performance compared to the IRLS estimator. Moreover, IRLS exhibited strong performance as the number of measurements increased. Finally, the ADMM-based prox-linear estimator required the fewest iterations for convergence, followed by the IRLS estimator and then the stochastic prox-linear estimator.

The third experiment aimed to assess the scalability and effectiveness of stochastic iterations on a larger power network consisting of 9,241 buses, which was available in the MATPOWER toolbox (Zimmerman et al. 2010). The true voltage magnitude for each bus was uniformly distributed within the range [0.95, 1.05], while the voltage angle ranged from $[-0.05\pi, 0.05\pi]$. The Gauss–Newton method was permitted a maximum of 10 iterations. In this experiment, all seven types of SCADA data were collected and subjected to additive noise, following the procedure from the previous experiment. This time, however, 5% of the data was compromised under model M2. The corrupted data $\xi_m := \tilde{\mathbf{v}}^{\mathrm{H}}\mathbf{H}_m\tilde{\mathbf{v}}$ relying on the individual $\mathbf{H}_m$ were generated using $\tilde{\mathbf{v}} \in \mathbb{R}^N$ from the standardized multivariate Gaussian distribution.

Table 2.3 Comparisons of the Gauss–Newton and stochastic LAV estimators

Algorithm	Number of iterations	Computation time (s)
Gauss–Newton Abur and Exposito (2004)	10 (maximum)	$\geq$ 3 hs
Stochastic LAV	22	2595 s

To put it simply, in this context, there were a grand total of 18,481 variables that needed to be estimated, and we gathered a comprehensive set of 91,919 measurements. Regrettably, 4.595 of these measurements were intentionally tampered with by adversaries. When initialized with the flat voltage profile point, the WLS-based Gauss–Newton iterations provided an estimate with a RMSE of 0.9846. In contrast, the stochastic scheme outlined in Algorithm 2.2, utilizing a diminishing stepsize $\mu_t = 100/t^{0.8}$, achieved a significantly improved RMSE of 0.0412. The computational runtimes for each of these schemes have been meticulously recorded in Table 2.3. It is unequivocally apparent that the stochastic LAV implementation significantly outpaces the computational demands of the WLS-based Gauss–Newton iterations, often completing the task in a fraction of the time.

Remark 2.3 Every Gauss–Newton iteration involves the computationally intensive task of inverting a matrix sized $(2N - 1) \times (2N - 1)$, thereby incurring a substantial computational complexity of $\mathcal{O}((2N - 1)^3)$. It is eminently clear that as the system size N grows, especially when $N \geq 10,000$, this cubic complexity of Gauss–Newton iterations, in addition to the substantial memory requirements, swiftly transforms into a computational bottleneck, even for a reasonably powerful desktop computer. In contrast, the per-iteration computational complexity of the proposed stochastic LAV scheme can be as minimal as $\mathcal{O}(1)$, a characteristic that undoubtedly scales seamlessly. This adaptability makes it ideally suited for tackling PSSE tasks of considerable dimensions. Consequently, it is intuitively apparent that in the realm of large-scale power systems, the proposed stochastic iterations-based LAV implementation vastly outperforms Gauss–Newton iterations in terms of computational speed and memory efficiency. Furthermore, this advantage of employing cost-effective stochastic iterations to grapple with large-scale optimization challenges has found resonance in recent breakthroughs in deep learning, particularly in domains such as visual recognition and speech translation. In these fields, stochastic gradient-based approaches, exemplified by stochastic gradient descent (SGD), have emerged as the cornerstone for training DNNs (LeCun et al. 2015). The success of these methods underscores the broader applicability and efficacy of stochastic iterations in tackling computationally demanding tasks.

Chapter 3
State Estimation from Rank-One Measurements

As utilities increasingly embrace smart grid technology and implement various upgrades, they expose themselves to inherent cyber vulnerabilities. Consequently, there has been a resurgence of interest in the field of robust PSSE. To address the challenges posed by malicious data, researchers have incorporated the LNR test into the PSSE process (Merrill and Schweppe 1971). Additionally, alternatives based on the least-median-squares and least-trimmed squares methods have been explored (Mili et al. 1996a). However, it is regrettable that these robust PSSE proposals often come with unfavorable computational complexities and/or stringent storage requirements, limiting their practical applicability in real-world power networks.

Conversely, the ℓ_1-based criterion is renowned in the domains of optimization and statistics for its robustness to outliers (Huber 2004, Abur and Exposito 2004). Beyond its robustness, the ℓ_1-based estimator also achieves statistical optimality in the maximum likelihood sense, particularly when dealing with independent additive noise following a Laplacian distribution. In the realm of PSSE literature, the ℓ_1-based criterion has been promoted for the identification and rejection of bad data (Kotiuga and Vidyasagar 1982). Recent research efforts have shifted toward the development of efficient and user-friendly algorithms tailored to address the non-convexity and non-smoothness challenges associated with the ℓ_1 loss function; examples of such efforts can be found in (Abur and Celik 1991; Jabr and Pal 2003).

In a noise-free environment, PSSE is equivalent to solving a quadratic equation system closely intertwined with power flow analysis, which is inherently complex. Assuming access to all power flow and voltage magnitude measurements, this chapter introduces an efficient algebraic technique. This technique transforms power flows into primary measurements, simplifying deviation minimization using the ℓ_1-based approach. To effectively tackle the inherent challenges of non-convexity and non-smoothness in ℓ_1-based PSSE, we present a deterministic approach rooted in composite optimization, which approximates linearity. We also explore its adaptation under stochastic gradient conditions. Furthermore, we establish criteria based on the ℓ_1-based loss function to ensure the precise retrieval and quadratic convergence

G. Wang et al., *Data-driven Optimization and Control for Autonomous Energy Systems*,
https://doi.org/10.1007/978-981-95-1782-4_3

properties of the proposed methodology. Simplified tests on various IEEE benchmark systems, conducted under diverse conditions, support our theoretical findings and demonstrate the rapid convergence and resilience of our proposed methods.

3.1 System Modeling and Problem Formulation

3.1.1 System Modeling

Consider an electric power grid, which we model as a graph denoted as $\mathcal{G} = (\mathcal{N}, \mathcal{L})$. In this graph, the nodes $\mathcal{N} := \{1, 2, \ldots, N\}$ correspond to individual buses, while the edges $\mathcal{L} := \{(n, n')\} \subseteq \mathcal{N} \times \mathcal{N}$ represent the transmission lines connecting these buses. Throughout the entirety of this section, our analyses are conducted within the framework of the per unit system. The complex voltage associated with each bus $n \in \mathcal{N}$ can be represented in rectangular coordinates as $v_n = \Re(v_n) + j\Im(v_n)$. For the sake of brevity, all nodal voltages are collectively organized into a vector $\mathbf{v} := [v_1 \cdots v_N]^{\mathrm{T}} \in \mathbb{C}^N$. Within the context of AC-based SE literature, the SCADA system (Abur and Exposito 2004) is typically capable of measuring a subset of the following system variables:

(a) $|v_n|$: the voltage magnitude at bus n;
(b) $P_{nn'}(Q_{nn'})n$ to: the active (reactive) power flow from buses $\{n'\}$ at the sending terminal;
(c) $P_n(Q_n)$: the active (reactive) power injection into bus n.

Consistent with the AC power flow model (Abur and Exposito 2004), these system variables can be mathematically formulated as quadratic functions of the voltage vector $\mathbf{v}$. This rationalizes the terminology referring to $\mathbf{v}$ as the system state. To this end, observe that the squared voltage magnitudes $V_n := |v_n|^2 = [\Re(v_n)]^2 + [\Im(v_n)]^2$ can be written as

$$V_n = \mathbf{v}^{\mathrm{H}}\mathbf{H}_n^V\mathbf{v}, \text{ with } \mathbf{H}_n^V := \mathbf{h}_n\mathbf{h}_n^{\mathrm{T}}, \tag{3.1}$$

for all $n \in \mathcal{N}$, where we have introduced the measurement vector $\mathbf{h}_n := \mathbf{e}_n$ with $\mathbf{e}_n$ being the n-th canonical vector in $\mathbb{R}^N$. To represent power injections as functions of the voltage vector $\mathbf{v}$, we introduce the bus admittance matrix $\mathbf{Y} = \mathbf{G} + j\mathbf{B} \in \mathbb{C}^N$, where $\mathbf{G}$ and $\mathbf{B} \in \mathbb{R}^{N\times N}$ represent the real and imaginary components of $\mathbf{Y}$ (Abur and Exposito 2004), respectively. In rectangular coordinates, the mathematical expressions for the active and reactive powers P_n and Q_n injected into bus n are as follows

$$P_n = \Re(v_n)\sum_{n'=1}^{N}[\Re(v_{n'})G_{nn'} - \Im(v_n)B_{nn'}] + \Im(v_n)\sum_{n'=1}^{N}[\Im(v_n)G_{nn'} + \Re(v_n)B_{nn'}], \tag{3.2}$$

$$Q_n = \Im(v_n)\sum_{n'=1}^{N}[\Re(v_n)G_{nn'} - \Im(v_n)B_{nn'}] - \Re(v_n)\sum_{n'=1}^{N}[\Im(v_n)G_{nn'} + \Re(v_n)B_{nn'}], \tag{3.3}$$

which can be compactly expressed as

$$P_n = \mathbf{v}^{\mathrm{H}}\mathbf{H}_n^P\mathbf{v}, \text{ with } \mathbf{H}_n^P := \frac{\mathbf{Y}_n^{\mathrm{H}} + \mathbf{Y}_n}{2}, \tag{3.4a}$$

$$Q_n = \mathbf{v}^{\mathrm{H}}\mathbf{H}_n^Q\mathbf{v}, \text{ with } \mathbf{H}_n^Q := \frac{\mathbf{Y}_n^{\mathrm{H}} - \mathbf{Y}_n}{2j}, \tag{3.4b}$$

with $\mathbf{Y}_n := \mathbf{e}_n\mathbf{e}_n^{\mathrm{T}}\mathbf{Y}$ for all $n \in \mathcal{N}$.

Concerning power flows, Kirchhoff's current law dictates that the complex current over the line $(n,\ n')$ at the 'sending' end is given by $i_{nn'} = y^s_{nn'}v_n + y_{nn'}(v_n - v_{n'})$, where $y^s_{nn'}$ represents the shunt admittance at bus n' associated with the line $(n,\ n')$. In accordance with the AC power flow model and Ohm's law, it is further asserted that the complex power flowing over line $(n,\ n')$ at the 'sending' end can be mathematically expressed as

$$S_{nn'} = P_{nn'} + jQ_{nn'} = v_n\bar{i}_{nn'} = |v_n|^2(\bar{y}^s_{nn'} + \bar{y}_{nn'}) - v_n\bar{v}_n\bar{y}_{nn'}. \tag{3.5}$$

It is worth pointing out that the complex power flow over line $(n,\ n') \in \mathcal{L}$ at the 'receiving' end is captured by that over line $(n',\ n) \in \mathcal{L}$ at the 'sending' end. Upon defining the following matrices for all lines $(n,\ n') \in \mathcal{L}$

$$\mathbf{Y}_{nn'} := (y^s_{nn'} + y_{nn'})\mathbf{e}_n\mathbf{e}_n^{\mathrm{T}} - \bar{y}_{nn'}\mathbf{e}_{n'}\mathbf{e}_{n'}^{\mathrm{T}}, \tag{3.6}$$

the active and reactive power flows at the 'sending' terminal, which correspond to the real and imaginary parts of $S_{nn'}$ in (3.5), can be succinctly represented as follows

$$P_{nn'} = \mathbf{v}^{\mathrm{H}}\mathbf{H}_{nn'}^P\mathbf{v}, \text{ with } \mathbf{H}_{nn'}^P := \frac{\mathbf{Y}_{nn'}^{\mathrm{H}} + \mathbf{Y}_{nn'}}{2}, \tag{3.7a}$$

$$Q_{nn'} = \mathbf{v}^{\mathrm{H}}\mathbf{H}_{nn'}^Q\mathbf{v}, \text{ with } \mathbf{H}_{nn'}^Q := \frac{\mathbf{Y}_{nn'}^{\mathrm{H}} - \mathbf{Y}_{nn'}}{2j}. \tag{3.7b}$$

Now that we have represented all SCADA measurements as functions of $\mathbf{v}$, the PSSE problem can be presented next.

3.1.2 Power System State Estimation

In practical cases, the SCADA system typically measures a subset of the system variables as defined in Eqs. (3.1), (3.4a), (3.4b), (3.7a) and (3.7b). Let's assume that a total of M such variables are measured, and these measurements are organized into a $M \times 1$ measurement vector as follows

$$\mathbf{z} := [\{\breve{V}_n\}_{n\in\mathcal{N}_V}, \{\breve{P}_n\}_{n\in\mathcal{N}_P}, \{\breve{Q}_n\}_{n\in\mathcal{N}_Q}, \{\breve{P}_{nn'}\}_{(n,n')\in\mathcal{L}_P}, \{\breve{Q}_{nn'}\}_{(n,n')\in\mathcal{L}_Q}]^{\mathrm{T}} \in \mathbb{R}^M, \tag{3.8}$$

where the check-marked terms represent possibly noisy observations of the corresponding error-free variables. The subsets $\mathcal{N}_V, \mathcal{N}_P, \mathcal{N}_Q \subseteq \mathcal{N}$, and $\mathcal{L}_P, \mathcal{L}_Q \subseteq \mathcal{L}$ specify the locations where meters are installed and the associated type of variables that are measured. In a more concise manner, per m-th measurement in $\mathbf{z}$ can be equivalently expressed as

$$z_m := \mathbf{v}^{\mathrm{H}}\mathbf{H}_m\mathbf{v} + \epsilon_m \tag{3.9}$$

for all $m \in \{1, 2, \ldots, M\}$, where the terms $\epsilon_m \in \mathbb{R}$ capture the metering errors and modeling inaccuracies, and the Hermitian measurement matrices $\mathbf{H}_m \in \mathbb{C}^{N\times N}$ are amenable to representing any subset of the matrices as defined in (3.1), (3.4a), (3.4b), (3.7a) and (3.7b). In the realm of PSSE, the paramount objective is the determination of $\mathbf{v} \in \mathbb{C}^N$ based upon the available data $(z_m; , \mathbf{H}_m)_{m=1}^M$.

Without loss of generality, adopting the least squares (LS) error objective, which coincides with the maximum likelihood criterion assuming additive white Gaussian noise, PSSE pursues problem

$$\min_{\mathbf{x}\in\mathbb{C}^N} \ell(\mathbf{x}) := \frac{1}{2M}\sum_{m=1}^{M}\left(z_m - \mathbf{x}^{\mathrm{H}}\mathbf{H}_m\mathbf{x}\right)^2. \tag{3.10}$$

Due to the quadratic terms within the squares, the quartic function $\ell(\mathbf{x})$ is nonconvex, with its general form being proven to be NP-hard (Pardalos and Vavasis 1991). Consequently, the computation of the LS or maximum likelihood estimate for $\mathbf{v}$ becomes computationally intractable in the general case.

3.1.3 Prior Contributions

The primary approach for minimizing the nonlinear LS loss, as presented in (3.10), is the Gauss–Newton method, widely acknowledged as the cornerstone in power system estimation (Abur and Exposito 2004). The Gauss–Newton method operates by linearizing all quadratic terms $\mathbf{x}^{\mathrm{H}}\mathbf{H}_m\mathbf{x}$ around an initial point using Taylor's expansion. It subsequently approximates the nonlinear LS fit defined in (3.10) with a linear one during per iteration. The method relies on the minimizer of this linear approximation to compute the next iterate (Wang et al. 2019c). Typically, the Gauss–Newton method converges rapidly, often requiring only a few iterations, especially for small- to medium-sized problems. Nonetheless, it is well-documented that Gauss–Newton iterations applied to non-convex LS problems exhibit sensitivity to the initial point and may encounter divergence in specific cases, as discussed in (Bertsekas 1999).

Conversely, numerous polynomial-time SE algorithms have been developed, focusing on convex programming techniques, as highlighted in (Zhu and Giannakis 2014). These convex methodologies leverage matrix lifting to represent all quadratic measurements $\mathbf{x}^{\mathrm{H}}\mathbf{H}_m\mathbf{x}$ as linear functions $\mathrm{Tr}(\mathbf{H}_m\mathbf{X})$, where $\mathbf{X} := \mathbf{x}\mathbf{x}^{\mathrm{H}} \succeq \mathbf{0}$ is a rank-one matrix variable. Upon relaxing the non-convex rank constraint, the originally nonlinear LS problem in (3.10) is reduced to, or can be equivalently reformulated as, a convex SDP. Concerning computational efficiency, these convex approaches involve the solution of an $N \times N$ positive semi-definite matrix within the context of M SDP constraints. The worst-case computational complexity associated with this task is $\mathcal{O}(M^4 N^{1/2} \log(1/\epsilon))$, where $\epsilon > 0$ denotes the desired solution accuracy (Ben-Tal and Nemirovski 2001). This complexity and the resultant storage requirement evidently do not scale nicely to the increasingly interconnected large power networks.

3.2 The Rank-One Measurement Approach

This section introduces an innovative framework for PSSE that capitalizes on recent developments in non-convex optimization. Specifically, our proposed approach addresses scalability, accuracy, and robustness by redefining the rank-two power flows in (3.7a) and (3.7b) as rank-one quadratic measurements. These measurements are characterized by corresponding measurement matrices with a rank of one. We then present two highly efficient algorithms designed to minimize the ℓ_1-based misfit of these transformed measurements. Unlike previous SE methods that involve the minimization of the quadratic polynomial in (3.10), our approach introduces a novel quadratic objective functional that is subsequently subjected to minimization. This shift in the objective function is a key feature of our methodology.

By employing more intricate algebraic transformations, our proposed framework can accommodate the inclusion of power injections. Furthermore, in the presence of additive noise, we provide numerical evidence showcasing the near-optimal statistical performance of the developed approach.

3.2.1 Measurement Transformation

In this section, our primary focus is on two types of measurements: (i) $\{|v_m|\}_{m=1}^N$, representing the voltage magnitudes at all buses, and ii) $\{P_{nn'}\}_{(m,n)\in\mathcal{L}_P}$ and/or $\{Q_{nn'}\}_{(m,n)\in\mathcal{L}_Q}$, indicating the active and/or reactive power flows on a selected subset of lines, denoted as $\mathcal{L}_P$, $\mathcal{L}_Q \subseteq \mathcal{L}$. Let's first consider the noise-free case, where all available measurements can be expressed as

$$z_m = \mathbf{v}^{\mathrm{H}}\mathbf{H}_m\mathbf{v}, \quad \forall m = 1, \ldots, M. \tag{3.11}$$

Without loss of generality, we can assume that the first N measurements correspond to the squared voltage magnitudes at the N buses, i.e., $z_m = |v_m|^2$ for $m = 1, 2, \ldots, N$, while the remaining measurements pertain to the (active or reactive) power flows. It is evident from Eq. (3.1) that the squared voltage magnitude measurements are defined as

$$z_m = \left|\mathbf{h}_m^{\mathrm{H}}\mathbf{v}\right|^2, \quad \forall m = 1, \ldots, N, \tag{3.12}$$

whose corresponding measurement matrices have rank one, that is, $\{\mathbf{H}_m = \mathbf{h}_m\mathbf{h}_m^{\mathrm{H}}\}_{m=1}^N$.

Now let us consider the power flow data pairs $(P_{nn'}; \mathbf{H}_{nn'}^P)$ and $(Q_{nn'}; \mathbf{H}_{nn'}^Q)$ in (3.7a) and (3.7b). Upon substituting $\mathbf{Y}_{nn'}$ in (3.6) into (3.7a) and (3.7b), one can rewrite for all lines $(n, n') \in \mathcal{L}_P$

$$\mathbf{H}_{nn'}^P = \frac{1}{2}\left(\alpha_{nn'}^P\mathbf{e}_n\mathbf{e}_n^{\mathrm{T}} - \beta_{nn'}^P\mathbf{e}_n\mathbf{e}_{n'}^{\mathrm{T}} - \overline{\beta}_{nn'}^P\mathbf{e}_{n'}\mathbf{e}_n^{\mathrm{T}}\right), \tag{3.13}$$

and similarly for all lines $(n, n') \in \mathcal{L}_Q$

$$\mathbf{H}_{nn'}^Q = \frac{1}{2}\left(\alpha_{nn'}^Q\mathbf{e}_n\mathbf{e}_n^{\mathrm{T}} - \beta_{nn'}^Q\mathbf{e}_n\mathbf{e}_{n'}^{\mathrm{T}} - \overline{\beta}_{nn'}^Q\mathbf{e}_{n'}\mathbf{e}_n^{\mathrm{T}}\right), \tag{3.14}$$

where the four coefficients are given by

$$\alpha_{nn'}^P := 2\Re(y_{nn'}^s + y_{nn'}), \quad \beta_{nn'}^P := y_{nn'}, \tag{3.15a}$$

$$\alpha_{nn'}^Q := -2\Im(y_{nn'}^s + y_{nn'}), \quad \beta_{nn'}^Q := jy_{nn'}. \tag{3.15b}$$

It is worth pointing out that $\alpha_{nn'}^P$ and $\alpha_{nn'}^Q$ are both real, while $\beta_{nn'}^P$ and $\beta_{nn'}^Q$ are in general complex.

As evident from (3.13) and (3.14), it is clear that each of the measurement matrices for active or reactive power flow, denoted as $\{\mathbf{H}_{nn'}^P\}_{(n,n')\in\mathcal{L}_P}$ or $\{\mathbf{H}_{nn'}^Q\}_{(n,n')\in\mathcal{L}_Q}$ is of at most rank-two. These matrices exhibit three nonzero entries, located at the (n, n)-, (n, n')-, and (n', n)-th positions, depending on the nonzero coefficients involved.

However, it should be noted that these power flow matrices are generally indefinite. Upon neglecting the superscripts for notational brevity, both $\mathbf{H}_{nn'}^{P}$ and $\mathbf{H}_{nn'}^{Q}$ are in the following form

$$\mathbf{H}_{nn'} = \frac{1}{2}\left(\alpha_{nn'}\mathbf{e}_n\mathbf{e}_n^{\mathrm{T}} - \beta_{nn'}\mathbf{e}_n\mathbf{e}_{n'}^{\mathrm{T}} - \overline{\beta}_{nn'}\mathbf{e}_{n'}\mathbf{e}_n^{\mathrm{T}}\right). \tag{3.16}$$

We establish the following result for power flow data $\{z_m\}$.

Proposition 3.1 (Rank-one measurements) *Suppose that the voltage magnitudes are measured at all buses. Then one can construct equivalently a new measurement* $\breve{z}_m$ *for each power flow* z_m *so that its corresponding measurement matrix* $\breve{\mathbf{H}}_{nn'}$ *is of rank-one, and positive semidefinite.*

Proof Depending on whether the coefficient $\alpha_{nn'}$ or $\beta_{nn'}$ is zero or not, we discuss separately the ensuing three cases:

(c1) $\alpha_{nn'} \neq 0$, $\beta_{nn'} \neq 0$;
(c2) $\alpha_{nn'} = 0$, $\beta_{nn'} \neq 0$;
(c3) $\alpha_{nn'} \neq 0$, $\beta_{nn'} = 0$,

where the trivial case of $\alpha_{nn'} = \beta_{nn'} = 0$ is excluded because all SCADA measurements are assumed nonzero. The objective of the subsequent section is to convert all power flow measurement matrices into rank-one (symmetric) positive semi-definite matrices. These three cases will be discussed individually in the following sections.

Consider first the case c1). When both $\alpha_{nn'}$ and $\beta_{nn'}$ are nonzero, the matrix $\mathbf{H}_{nn'}$ has exactly rank two with three nonzero entries. Its two nonzero eigenvalues can be obtained by solving the quadratic equation

$$\lambda\left(\lambda - \frac{\alpha_{nn'}}{2}\right) - \frac{|\beta_{nn'}|^2}{4} = 0, \tag{3.17}$$

which is derived by setting the determinant of $(\lambda\mathbf{I}_N - \mathbf{H}_{nn'})$ to zero. Its closed-form solutions are given by

$$\lambda_1 = \frac{\alpha_{nn'} + \sqrt{\alpha_{nn'}^2 + 4|\beta_{nn'}|^2}}{4} > 0,$$
$$\lambda_2 = \frac{\alpha_{nn'} - \sqrt{\alpha_{nn'}^2 + 4|\beta_{nn'}|^2}}{4} < 0.$$

Let $\mathbf{u}_1, \mathbf{u}_2 \in \mathbb{C}^N$ be the unit eigenvectors of $\mathbf{H}_{nn'}$ associated with the eigenvalues λ_1, λ_2, respectively. Hence, one can write $\mathbf{H}_{nn'} := \lambda_1\mathbf{u}_1\mathbf{u}_1^{\mathrm{H}} + \lambda_2\mathbf{u}_2\mathbf{u}_2^{\mathrm{H}}$. To obtain a rank-one positive semi-definite matrix, the first attempt would be to compensate for the negative eigenvalue λ_2 and make it zero. This is tantamount to adding $-\lambda_2\mathbf{u}_2\mathbf{u}_2^{\mathrm{H}}$ to $\mathbf{H}_{nn'}$, and accordingly adding $-\lambda_2\mathbf{v}^{\mathrm{H}}(\mathbf{u}_2\mathbf{u}_2^{\mathrm{H}})\mathbf{v}$ to the measurement $z_{nn'}$, that is

$$\mathbf{H}_{nn'} := \mathbf{H}_{nn'} - \lambda_2 \mathbf{u}_2 \mathbf{u}_2^{\mathrm{H}}, \tag{3.18a}$$

$$\breve{z}_{nn'} := z_{nn'} - \lambda_2 \mathbf{v}^{\mathrm{H}} (\mathbf{u}_2 \mathbf{u}_2^{\mathrm{H}}) \mathbf{v}. \tag{3.18b}$$

In this case, the transformed measurement matrix $\breve{\mathbf{H}}_{nn'}$ becomes rank-one and symmetric positive semi-definite, resulting in $\breve{z}_{nn'}$ as the transformed measurement. However, achieving this transformation involves computing the term $\mathbf{v}^{\mathrm{H}}(\mathbf{u}_2\mathbf{u}_2^{\mathrm{H}})\mathbf{v}$, which necessitates knowledge of the true state vector $\mathbf{v}$. Unfortunately, this procedure is not practically feasible, and an innovative approach is required to overcome this obstacle.

Recalling our working assumption of having access to all squared voltage magnitudes $\{|v_n|^2\}_{n=1}^N$, we demonstrate in the following that it is possible to enhance $\mathbf{H}_{nn'}$ by adding a matrix of the form $(\delta_{nn'}/2)\mathbf{e}_{n'}\mathbf{e}_{n'}^{\mathrm{T}}$ to it. This augmentation results in the sum, denoted as

$$\breve{\mathbf{H}}_{nn'} := \mathbf{H}_{nn'} + (\delta_{nn'}/2)\mathbf{e}_{n'}\mathbf{e}_{n'}^{\mathrm{T}}, \tag{3.19}$$

can be rendered rank-one. To determine the unknown coefficient $\delta_{nn'}$, we proceed by setting the determinant of $(\lambda \mathbf{I}_N - \breve{\mathbf{H}}_{nn'})$ to zero, which results in the following equation

$$\left(\lambda - \frac{\alpha_{nn'}}{2}\right)\left(\lambda - \frac{\delta_{nn'}}{2}\right) - \frac{|\beta_{nn'}|^2}{4} = 0. \tag{3.20}$$

To ensure the rank-one property of the matrix $\breve{\mathbf{H}}_{nn'}$, we aim for the quadratic Eq. (3.20) to possess only one nonzero solution. In basic linear algebra terms, achieving this condition is equivalent to having a zero constant term in (3.20), which leads to the following equation

$$\alpha_{nn'}\delta_{nn'} - |\beta_{nn'}|^2 = 0,$$

or alternatively,

$$\delta_{nn'} := |\beta_{nn'}|^2/\alpha_{nn'}.$$

It can be verified that the transformed measurement matrix

$$\breve{\mathbf{H}}_{nn'} := \mathbf{H}_{nn'} + (|\beta_{nn'}|^2/(2\alpha_{nn'}))\mathbf{e}_{n'}\mathbf{e}_{n'}^{\mathrm{T}} \tag{3.21}$$

is rank-one. In addition, if $\alpha_{nn'} > 0$, then $\breve{\mathbf{H}}_{nn'}$ is positive semi-definite. Therefore, one can write

$$\breve{\mathbf{H}}_{nn'} := \mathbf{h}_{nn'}\mathbf{h}_{nn'}^{\mathrm{H}} \tag{3.22}$$

with the equivalent measurement vector being

$$\mathbf{h}_{nn'} := \sqrt{\frac{\alpha_{nn'}}{2}}\mathbf{e}_n + \frac{\overline{\beta}_{nn'}}{\sqrt{2\alpha_{nn'}}}\mathbf{e}_{n'}. \tag{3.23}$$

The transformed measurement $\breve{z}_{nn'}$ for $\breve{\mathbf{H}}_{nn'}$ can be expressed as

$$\breve{z}_{nn'} := \mathbf{v}^{\mathrm{H}}\breve{\mathbf{H}}_{nn'} = \mathbf{v}^{\mathrm{H}}\mathbf{H}_{nn'}\mathbf{v} + \mathbf{v}^{\mathrm{H}}\big(\delta_{nn'}\mathbf{e}_{n'}\mathbf{e}_{n'}^{\mathrm{T}}/2\big)\mathbf{v} = z_{nn'} + (|\beta_{nn'}|^2/2\alpha_{nn'})|v_{n'}|^2, \tag{3.24}$$

for which the required quantity $(|\beta_{nn'}|^2/2\alpha_{nn'})|v_{n'}|^2$ is available, or can be obtained as long as $|v_n|^2$ is available.

If $\alpha_{nn'} < 0$, one can instead define

$$\breve{\mathbf{H}}_{nn'} := -\mathbf{H}_{nn'} + (|\beta_{nn'}|^2/(2\alpha_{nn'}))\mathbf{e}_{n'}\mathbf{e}_{n'}^{\mathrm{T}}, \tag{3.25}$$

and express the corresponding measurement vector as

$$\mathbf{h}_{nn'} := \sqrt{\frac{-\alpha_{nn'}}{2}}\mathbf{e}_n + \frac{\overline{\beta}_{nn'}}{\sqrt{-2\alpha_{nn'}}}\mathbf{e}_{n'}, \tag{3.26}$$

for which the corresponding measurement becomes

$$\breve{z}_{nn'} := \mathbf{v}^{\mathrm{H}}\breve{\mathbf{H}}_{nn'}\mathbf{v} = -\mathbf{v}^{\mathrm{H}}\mathbf{H}_{nn'}\mathbf{v} - \mathbf{v}^{\mathrm{H}}\big(\delta_{nn'}\mathbf{e}_{n'}\mathbf{e}_{n'}^{\mathrm{T}}/2\big)\mathbf{v} = -z_{nn'} - (|\beta_{nn'}|^2/2\alpha_{nn'})|v_n|^2. \tag{3.27}$$

Similarly, the quantity $-(|\beta_{nn'}|^2/2\alpha_{nn'})|v_n|^2$ is also accessible under our working assumption.

Let us turn our attention to case c2). To convert $\mathbf{H}_{nn'}$ into a rank-one positive semi-definite matrix, we can achieve this by adding a matrix of the form $(\gamma_{nn'}/2)\mathbf{e}_n\mathbf{e}_n^{\mathrm{T}} + (\delta_{nn'}/2)\mathbf{e}_{n'}\mathbf{e}_{n'}^{\mathrm{T}}$, resulting in the following transformation

$$\breve{\mathbf{H}}_{nn'} := \mathbf{H}_{nn'} + (\gamma_{nn'}/2)\mathbf{e}_n\mathbf{e}_n^{\mathrm{T}} + (\delta_{nn'}/2)\mathbf{e}_{n'}\mathbf{e}_{n'}^{\mathrm{T}} \tag{3.28}$$

for some coefficients $\gamma_{nn'} > 0$ and $\delta_{nn'} > 0$ to be determined. Similar to the discussion for case c1), to find $\gamma_{nn'}$ and $\delta_{nn'}$, one sets the determinant of $(\lambda\mathbf{I}_N - \breve{\mathbf{H}}_{nn'})$ to zero, leading to $\big(\lambda - \frac{\gamma_{nn'}}{2}\big)\Big(\lambda - \frac{\delta_{nn'}}{2}\Big) - \frac{|\beta_{nn'}|^2}{4} = 0$.

The fact that $\breve{\mathbf{H}}_{nn'}$ is rank-one implies that

$$\gamma_{nn'}\delta_{nn'} - |\beta_{nn'}|^2 = 0.$$

Without sacrificing generality, it is permissible to adopt that

$$\gamma_{nn'} := 1 \text{ and } \delta_{nn'} := |\beta_{nn'}|^2,$$

and in accordance with Eq. (3.28), the matrix $\breve{\mathbf{H}}_{nn'}$ becomes rank-one, and it can be expressed as

$$\breve{\mathbf{H}}_{nn'} := \mathbf{h}_{nn'}\mathbf{h}_{nn'}^{\mathrm{H}} \tag{3.29}$$

with

$$\mathbf{h}_{nn'} := \frac{1}{\sqrt{2}}\mathbf{e}_n - \frac{\beta_{nn'}}{\sqrt{2}}\mathbf{e}_n. \tag{3.30}$$

The transformed measurement associated with $\breve{\mathbf{H}}_{nn'}$ can be computed as follows

$$\begin{aligned} \breve{z}_{nn'} := \mathbf{v}^{\mathrm{H}}\mathbf{H}_{nn'}\mathbf{v} &= \mathbf{v}^{\mathrm{H}}\mathbf{H}_{nn'}\mathbf{v} + \mathbf{v}^{\mathrm{H}}\big(\beta_{nn'}\mathbf{e}_n\mathbf{e}_n^{\mathrm{T}}/2 + \delta_{nn'}\mathbf{e}_{n'}\mathbf{e}_{n'}/2\big)\mathbf{v} \\ &= z_{nn'} + (1/2)|v_n|^2 + (|\beta_{nn'}|^2/2)|v_{n'}|^2, \end{aligned} \tag{3.31}$$

for which the required quantities $|v_n|^2$ and $|\beta_{nn'}|^2|v_{n'}|^2$ can be computed when $|v_n|^2$ and $|v_{n'}|^2$ are available.

Now, let's consider the final case c3). In this case, with $\beta_{nn'} = \overline{\beta}_{nn'} = 0$, we can represent $\mathbf{H}_{nn'}$ as $\alpha_{nn'}\mathbf{e}_n\mathbf{e}_n^{\mathrm{T}}/2$, which is inherently rank-one. To ensure its positive semi-definiteness, we simply take the absolute value and define

$$\breve{\mathbf{H}}_{nn'} := \mathbf{h}_{nn'}\mathbf{h}_{nn'}^{\mathrm{H}} = \left(\sqrt{\frac{|\alpha_{nn'}|}{2}}\mathbf{e}_n\right)\left(\sqrt{\frac{|\alpha_{nn'}|}{2}}\mathbf{e}_n\right)^{\mathrm{T}}. \tag{3.32}$$

The transformed measurement is given by

$$\breve{z}_{nn'} = |z_{nn'}|. \tag{3.33}$$

In summary, for any set of active or reactive power flow measurements $(z_{nn'}; \mathbf{H}_{nn'})$, we have devised a method to derive a new measurement pair $(\breve{z}_{nn'}; \breve{\mathbf{H}}_{nn'})$, wherein $\breve{\mathbf{H}}_{nn'}$ is transformed into a rank one, positive semi-definite matrix. This transformation is achieved through the steps outlined in (3.22–3.33), relying on the values of coefficients $\alpha_{nn'}$ and $\beta_{nn'}$, under the assumption that the voltage magnitudes at all buses are accessible.

The assumption on full voltage measurements can be relaxed. Indeed, it is possible to build up rank-one measurements via linear combinations, so long as two of the following SCADA quantities $\{|v_n|, |v_{n'}|, p_{nn'}, p_{n'n}, q_{nn'}, q_{n'n}\}$ are measured on every line $(n, n') \in \mathcal{L}$. In a nutshell, one can readily rewrite all measurements as intensities of some known and deterministic linear transforms of the state vector, namely

$$\breve{z}_m = \left|\mathbf{h}_m^{\mathrm{H}}\mathbf{v}\right|^2, \quad \forall m = 1, \ldots, M, \tag{3.34}$$

where the measurement vectors $\mathbf{h}_m \in \mathbb{C}^N$ are given in (3.1), (3.23), (3.26), (3.30), and (3.32), while the corresponding transformed measurements $\breve{z}_m > 0$ are defined in (3.1), (3.24), (3.27), (3.31), and (3.33). Additionally, it's noteworthy that all these vectors $\mathbf{h}_m$ exhibit high sparsity, with at most two nonzero entries, independent of the system size N. This inherent sparsity can be strategically leveraged to enhance the computational efficiency and scalability of the iterative PSSE solvers.

3.3 Prox-Linear State Estimation Solvers

In the general context of solving a set of consistent quadratic equations, it's important to note that there can be multiple solutions, even when trivial ambiguities are excluded. In the specific domain of phase retrieval, where measurement matrices take the form of rank-one positive semi-definite matrices (i.e., $\mathbf{H}_m = \mathbf{h}_m\mathbf{h}_m^{\mathrm{H}}$), it has been established that a minimum of $M \geq 4N - 4$ random quadratic equations is required to ensure the uniqueness of the solution (Hering et al. 2013). However, in the realm of power systems research, the question of how many quadratic measurements, as represented in Eqs. (3.1), (3.4a), (3.4b), (3.7a), and (3.7b), are needed for the unique determination of the PSSE solution remains an open challenge. To make practical progress in this contribution, we assume the availability of a sufficiently large number of measurements. We operate under the assumption that this collective set of measurements collectively determines a unique solution, representing the underlying true system state.

In this section, we harness the rank-one measurement model to introduce two algorithms that offer scalable and precise power system state recovery through non-convex optimization. Specifically, our focus shifts towards minimizing the ℓ_1 loss, tailored to match the intensity measurements $\breve{z}_m$ in (3.34), as opposed to the noiseless z_m in (3.11). Despite the inherent non-convexity and non-smoothness of the resulting loss function, we first present a deterministic prox-linear algorithm In cases where the initialization is sufficiently close to the true underlying voltage state vector, and the loss function adheres to specific local 'stability conditions', our initial deterministic approach showcases the ability to recover the genuine voltage vector at a quadratic convergence rate. This process involves solving a quadratic program with per iteration, a task for which readily available off-the-shelf convex programming toolboxes can be employed. However, it's essential to acknowledge that the computational complexity associated with this approach does not scale optimally with increasing system size. To address this scalability challenge, we pursue a stochastic extension of the algorithm This stochastic variant processes a single measurement per iteration, substantially enhancing the algorithm's scalability while maintaining accuracy and precision in PSSE.

It is a well-established principle in both statistics and power systems literature that ℓ_1-based loss functions lead to estimators based on medians. Moreover, they exhibit a robustness that allows them to handle significant errors in the measurements $\breve{z}_m$ in a relatively forgiving manner (Duchi and Ruan 2019). This prompts us to consider the ℓ_1 loss formulation

$$\min_{\mathbf{x}\in\mathbb{C}^N} \ell(\mathbf{x}) := \frac{1}{M}\sum_{m=1}^{M}\left|\breve{z}_m - |\mathbf{h}_m^{\mathrm{H}}\mathbf{x}|^2\right|. \tag{3.35}$$

Certainly, this loss function exhibits distinctive non-smooth and non-convex characteristics, with no local convexity present around $\mathbf{v}$. This behavior becomes evident when considering a straightforward scalar case example, $\ell(x) = |1 - x^2|$. Consequently, the efficient minimization of such functions presents a significant challenge.

However, this function possesses several intriguing structural properties that we can leverage to craft efficient iterative algorithms for solving the problem (3.35) effectively, as elaborated in previous works (Burke and Ferris 1995; Lewis and Wright 2016; Duchi and Ruan 2019). To illustrate this, let's start by expressing the loss function as a composition: $\ell(\mathbf{x}) := c(\mathbf{s}(\mathbf{x}))$. Here, $c(\cdot) := \|\cdot\|_1$ is a convex function, and $\mathbf{s}(\mathbf{x}) := \frac{1}{M}\left(\breve{\mathbf{z}} - |\mathbf{H}\mathbf{x}|^2\right)$ is a smooth one. In these expressions, $\breve{\mathbf{z}} := [\breve{z}_1 \cdots \breve{z}_M]^{\mathrm{T}}$, $\mathbf{H} := [\mathbf{h}_1 \cdots \mathbf{h}_M]^{\mathrm{H}}$, and the modulus operator $|\cdot|$ should be understood as element-wise when applied to a vector. This compositional structure lends itself well to iterative procedures known as proximal linear (prox-linear) algorithms, which we will delve into in the following sections.

Consider a real-valued $\mathbf{x} \in \mathbb{R}^N$ for now. In the context of the (deterministic) prox-linear method for minimizing $\ell(\mathbf{x}) = c(\mathbf{s}(\mathbf{x}))$, the strategy involves linearizing only $\mathbf{s}$ and then proceeding to minimize a series of locally regularized models. To elucidate this, let's commence with an initial point $\mathbf{x}_0 \in \mathbb{R}^N$ within our discussion. The prox-linear method, at per iteration with the current iterate $\mathbf{x}_t \in \mathbb{R}^N$, defines a local 'linearization' of ℓ as follows

$$\ell_{\mathbf{x}_t}(\mathbf{x}) := c\big(\mathbf{s}(\mathbf{x}_t) + \nabla^{\mathrm{T}}\mathbf{s}(\mathbf{x}_t)(\mathbf{x} - \mathbf{x}_t)\big), \tag{3.36}$$

where $\nabla\mathbf{s}(\mathbf{x}_t) \in \mathbb{R}^{N\times M}$ denotes the Jacobian matrix of $\mathbf{s}$ evaluated at point $\mathbf{x}_t$. The prox-linear method then proceeds iteratively to obtain subsequent iterates $\mathbf{x}_1$, $\mathbf{x}_2$, and so forth, by minimizing these quadratically regularized models, a technique well-established in literature (Burke and Ferris 1995; Lewis and Wright 2016).

$$\mathbf{x}_{t+1} = \arg\min_{\mathbf{x}\in\mathbb{R}^N} \left\{ \ell_{\mathbf{x}_t}(\mathbf{x}) + \frac{1}{2\mu_t}\|\mathbf{x} - \mathbf{x}_t\|_2^2 \right\}, \tag{3.37}$$

where $\mu_t > 0$ is a step size that can be fixed a priori to some constant, or be determined 'on-the-fly' through a line search h (Burke and Ferris 1995; Lewis and Wright 2016). Furthermore, it is worth noting that the linearization $\ell_{\mathbf{x}_t}(\mathbf{x})$ is convex with respect to $\mathbf{x}$, rendering problem (3.37) convex as well. Previous research, as outlined in (Lewis and Wright 2016), has demonstrated that when c is L-Lipschitz and $\nabla\mathbf{s}$ is κ-Lipschitz, selecting any step size $0 < \mu < \frac{1}{\kappa L}$ ensures that the algorithm (3.37) possesses two crucial properties: (i) is a descent method; that is, the iterates $\{\mathbf{x}_k\}$ monotonically decrease the function value of $\ell(\mathbf{x})$; and (ii) finds an (approximate) stationary point of (3.35).

However, it's essential to note that the PSSE problem (3.35) involves optimization over complex-valued variables, with $\mathbf{x} \in \mathbb{C}^N$. Upon examination, it becomes evident that the functions ℓ and $\mathbf{s}$ do not satisfy the Cauchy-Riemann equations; the definition of Cauchy-Riemann equations is provided in (Markouchevitch 1977). Consequently, functions ℓ and $\mathbf{s}$ are not holomorphic, meaning they are not complex-differentiable with respect to $\mathbf{x}$. Therefore, relying solely on the 'linearization' or the first-order Taylor's expansion of $\mathbf{s}(\mathbf{x})$ in $\mathbf{x} \in \mathbb{C}^N$ [see (3.36)] is not possible. To tackle this challenge, we turn to Wirtinger's calculus, a mathematical tool that extends prox-linear algorithms to handle optimization over complex-valued arguments. For foundational information on Wirtinger's calculus, please refer to (Brandwood 1983).

3.3.1 Deterministic Prox-Linear SE Solver

Our initial deterministic prox-linear approach to solving (3.35) can be succinctly described as follows: we commence with an initialization $\mathbf{x}_0 := \mathbf{1} \in \mathbb{R}^N$ and then systematically minimize quadratically regularized functions centered at the current iterate $\mathbf{x}_t \in \mathbb{C}^N$ to obtain the subsequent iterate

$$\mathbf{x}_{t+1} = \arg\min_{\mathbf{x}\in\mathbb{C}^N} \frac{1}{M}\sum_{m=1}^{M}\left|b_{m,t} - 2\Re\left(\mathbf{a}_{m,t}^G\mathbf{x}\right)\right| + \frac{1}{2\mu_t}\|\mathbf{x} - \mathbf{x}_t\|_2^2, \tag{3.38}$$

where the term $\{b_{m,t} - 2\Re(\mathbf{a}_{m,t}^G\mathbf{x})\}$ can be viewed as the first-order Taylor's approximation of the non-holomorphic function $\breve{z}_m - \left|\mathbf{h}_m^G\mathbf{x}\right|^2$ at the point $\mathbf{x}_t$, computed using the Wirtinger derivatives. This is evident in the subsequent derivation.

By introducing complex conjugate coordinates $[\mathbf{x}^\mathrm{T}\,\overline{\mathbf{x}}^\mathrm{T}]^\mathrm{T} \in \mathbb{C}^{2N}$, $\mathbf{s}(\mathbf{x}) = \mathbf{s}(\mathbf{x}, \overline{\mathbf{x}}) \in \mathbb{C}^M$. It is evident that $\mathbf{s}(\mathbf{x}, \overline{\mathbf{x}})$ becomes holomorphic in $\mathbf{x}$ for a fixed $\overline{\mathbf{x}}$, and vice versa. This gives rise to the concept of partial Wirtinger derivatives (Brandwood 1983), which can be expressed as follows

$$\frac{\partial s_m}{\partial \mathbf{x}} := \frac{\partial s_m(\mathbf{x}, \overline{\mathbf{x}})}{\partial \mathbf{x}}\bigg|_{\overline{\mathbf{x}}=\text{constant}} = \left[\frac{\partial s_m}{\partial x_1}\frac{\partial s_m}{\partial x_2}\cdots\frac{\partial s_m}{\partial x_N}\right],$$
$$\frac{\partial s_m}{\partial \overline{\mathbf{x}}} := \frac{\partial s_m(\mathbf{x}, \overline{\mathbf{x}})}{\partial \overline{\mathbf{x}}}\bigg|_{\mathbf{x}=\text{constant}} = \left[\frac{\partial s_m}{\partial \overline{x}_1}\frac{\partial s_m}{\partial \overline{x}_2}\cdots\frac{\partial s_m}{\partial \overline{x}_N}\right],$$

where the partial derivative with respect to $\mathbf{x}$ ($\overline{\mathbf{x}}$) treats $\overline{\mathbf{x}}$ ($\mathbf{x}$) as a constant in s_m.The complex gradient of $s_m(\mathbf{x}, \overline{\mathbf{x}})$ with respect to $\mathbf{x}$ or $\overline{\mathbf{x}}$ can be defined by

$$\overline{\mathbf{x}}\nabla_{\mathbf{x}} s_m := \left(\frac{\partial s_m}{\partial \mathbf{x}}\right)^{\mathrm{H}}, \text{ and } \nabla_{\overline{\mathbf{x}}} s_m := \left(\frac{\partial s_m}{\partial \overline{\mathbf{x}}}\right)^{\mathrm{H}},$$

yielding the complex gradient of s_m in new coordinate system

$$\nabla_c s_m := \left[\nabla_{\mathbf{x}}^{\mathrm{T}} s_m \nabla_{\overline{\mathbf{x}}}^{\mathrm{T}} s_m\right]^{\mathrm{T}} = \left[\frac{\partial s_m}{\partial \mathbf{x}}\frac{\partial s_m}{\partial \overline{\mathbf{x}}}\right]^{\mathrm{H}}.$$

Upon introducing the complex Jacobian

$$\nabla_c \mathbf{s} := [\nabla_c s_1 \nabla_c s_2 \cdots \nabla_c s_M] \in \mathbb{C}^{2N\times M},$$

we can define the first-order Taylor's expansion for given vectors $\mathbf{x}$ and $\Delta\mathbf{x} \in \mathbb{C}^N$ as follows

$$\mathbf{s}(\mathbf{x}+\Delta\mathbf{x}) \approx \mathbf{s}(\mathbf{x}) + \nabla_c^{\mathrm{H}}\mathbf{s}(\mathbf{x})\left[\frac{\Delta\mathbf{x}}{\overline{\Delta\mathbf{x}}}\right] = \mathbf{s}(\mathbf{x}) + 2\Re\left(\nabla^{\mathrm{H}}\mathbf{s}_{\mathbf{x}}(\mathbf{x})\Delta\mathbf{x}\right) \in \mathbb{R}^{M\times N}. \quad (3.39)$$

The coefficients $\mathbf{a}_{m,t}$ and $b_{m,t}$ are given by

$$\mathbf{a}_{m,t} := \left(\mathbf{h}_m^G \mathbf{x}_t\right)\mathbf{h}_m, \quad (3.40a)$$

$$b_{m,t} := \breve{z}_m + \left|\mathbf{h}_m^G \mathbf{x}_t\right|^2. \quad (3.40b)$$

Notice that the problem (3.38), which is addressed in per iteration of our deterministic prox-linear SE solver, is a convex quadratic program. This type of problem can be efficiently solved using standard convex programming techniques. Under appropriate conditions, our scheme converges quadratically fast to the true state vector $\mathbf{v}$, meaning that we have to solve only about $\log_2\log_2(1/\epsilon)$ such quadratic programs to obtain an estimate $\mathbf{x}$ of $\mathbf{v}$ satisfying $\text{dist}(\mathbf{x}, \mathbf{v}) \le \epsilon\|\mathbf{v}\|_2$. As demonstrated by our numerical experiments in Sect. 5.4, this typically translates to around 5 to 8 convex quadratic programs in real-world cases. Furthermore, our approach is versatile and applies effectively both in noiseless situations and when a constant (random) portion of the measurements may be corrupted.

For details of deterministic prox-linear solver, please refer to Algorithm 3.1. In terms of computational complexity, the data preparation in Step 2 can be executed in $O(M)$ operations. Leveraging the sparsity of $\mathbf{h}_m$'s, the computation of coefficients $\{(b_{m,t}, \mathbf{a}_{m,t})\}_{m=1}^{M}$ in Step 5 can also be accomplished with $O(M)$ operations. The overall computational complexity of Algorithm 3.1 is primarily determined by solving the quadratic program in (3.38) in Step 6. Using standard convex programming solvers, this typically results in a complexity of $O(MN^2)$. However, iterative approaches based on the ADMM can reduce this to $\mathfrak{O}(MN \log(1/\epsilon))$ (Duchi and Ruan 2019, Wang et al. 2019b). Nonetheless, the latter complexity may still pose challenges for large-scale power networks. Additionally, while $\{\mathbf{h}_m\}_{m=1}^{M}$ exhibit sparsity with at most two nonzero entries, this property cannot be fully leveraged to accelerate the computation of the quadratic program in (3.38). To address these concerns, we propose an alternative stochastic approach to (3.38) for solving the problem (3.35).

Algorithm 3.1 Deterministic Prox-linear SE Solve

1: **Input** data $\{(z_m, \mathbf{H}_m)\}_{m=1}^{M}$, step size $\mu > 0$, initialization $\mathbf{v}_0 \in \mathbb{C}^N$, solution accuracy $\epsilon > 0$, and set $t = 0$

2: **Prepare** the power flow data $\{(z_m, \mathbf{H}_m)\}_{m=N+1}^{M}$ according to (3.1), (3.23–3.24), (3.26–3.27), (3.30–3.31), and (3.32–3.33) to obtain $\{(\breve{z}_m, \mathbf{h}_m)\}_{m=N+1}^{M}$ based on $\{z_m = |v_m|^2\}_{m=1}^{N}$

3: **Repeat**

4: **Evaluate** $\mathbf{a}_{m,t}$ and $b_{m,t}$ in (3.40a) and (3.40b)

5: **Solve** (3.38) to yield $\mathbf{x}_{t+1}$

6: $t = t + 1$

7: **Until** $\|\mathbf{x}_t - \mathbf{x}_{t-1}\|_2 \leq \varepsilon\sqrt{N}$

8: **Return** $\mathbf{x}_t$

3.3.2 *Stochastic Prox-Linear SE Solver*

The stochastic prox-linear method processes one measurement per iteration. Starting with an initial guess $\mathbf{x}_0$, our stochastic prox-linear SE solver proceeds by selecting a loss function uniformly at random through the random choice of $m_t \in \{1, 2, \ldots, M\}$. It then minimizes the quadratically regularized 'linearization' of this selected loss function around the current estimate $\mathbf{x}_t$ to obtain $\mathbf{x}_{t+1}$ (Duchi and Ruan 2018). In other words, the iterative procedure is defined recursively for $t = 0, 1, 2, \ldots$ as follows

$$\mathbf{x}_{t+1} = \arg\min_{\mathbf{x} \in \mathbb{C}^N} \left| b_{m_t,t} - 2\Re\left(\mathbf{a}_{m_t,t}^{\mathrm{H}} \mathbf{x}\right) \right| + \frac{1}{2\mu_t} \|\mathbf{x} - \mathbf{x}_t\|_2^2, \tag{3.41}$$

where the coefficients $b_{m_t,t}$ and $\mathbf{a}_{m_t,t}$ are given in (3.40a) and (3.40b), with $b_{m_t,t} - 2\Re(\mathbf{a}_{m_t,t}^{\mathrm{H}}\mathbf{x})$ being the first-order Taylor's expansion of the m_t-th error function $\breve{z}_{m_t} - |\mathbf{h}_{m_t}^{\mathrm{H}}\mathbf{x}|^2$ around $\mathbf{x}_t$. Indeed, problem (3.41) also falls into the category of quadratic programs. However, in contrast to the first quadratic program in (3.37), the good news is that the solution to (3.41) can be readily derived in a simple closed form.

To that end, an earlier result in (Wang et al. 2019b) is invoked, as shown below for completeness.

Proposition 3.2 (Wang et al. 2019b, Prop. 3.3) *Given* $\mathbf{a} \in \mathbb{C}^N$ *and* $b \in \mathbb{R}$, *the solution of*

$$\min_{\mathbf{w}\in\mathbb{C}^N} \left|b - \Re\left(\mathbf{a}^{\mathrm{H}}\mathbf{w}\right)\right| + \frac{1}{2\mu}\|\mathbf{w}\|_2^2, \tag{3.42}$$

can be obtained as $\widehat{\mathbf{w}} := \mathrm{proj}_{\mu}(b/\|\mathbf{a}\|_2^2) \cdot \mathbf{a}$, *where* $\mathrm{proj}_{\mu}(x) : \mathbb{R} \times \mathbb{R}_+ \to \mathbb{R}$ *is the projection operator that returns the real number in interval* $[-\tau, \tau]$ *closest to any given* $x \in \mathbb{R}$.

Condition 1 (Duchi and Ruan 2019) For any given $\mathbf{v} \in \mathbb{R}^N$, there exists a parameter $\rho > 0$ such that function $\ell(\mathbf{x})$ satisfies the following for all $\mathbf{x} \in \mathbb{R}^n$

$$\ell(\mathbf{x}) - \ell(\mathbf{v}) \geq \rho\|\mathbf{x} - \mathbf{v}\|_2\|\mathbf{x} + \mathbf{v}\|_2. \tag{3.43}$$

Concerning the lower bound in (3.43), we have the next result.

Lemma 3.1 *For any fixed* $\mathbf{v} \in \mathbb{C}^N$, *the inequality holds for all* $\mathbf{x} \in \mathbb{C}^N$.

$$\|\mathbf{x} - \mathbf{v}\|_2^2\|\mathbf{x} + \mathbf{v}\|_2^2 - 4\left|\Im(\mathbf{x}^{\mathrm{H}}\mathbf{v})\right|^2 \geq \|\mathbf{v}\|_2^2\mathrm{dist}^2(\mathbf{x}, \mathbf{v}). \tag{3.44}$$

Proof The left-hand-side term of (3.44) can be rewritten as

$$\begin{aligned}
&\|\mathbf{x} - \mathbf{v}\|_2^2\|\mathbf{x} + \mathbf{v}\|_2^2 - 4\Im^2(\mathbf{x}^{\mathrm{H}}\mathbf{v}) \\
&= \left(\|\mathbf{x}\|_2^2 + \|\mathbf{v}\|_2^2 - 2\Re(\mathbf{x}^{\mathrm{H}}\mathbf{v})\right)\left(\|\mathbf{x}\|_2^2 + \|\mathbf{v}\|_2^2 + 2\Re(\mathbf{x}^{\mathrm{H}}\mathbf{v})\right) - 4\Im^2(\mathbf{x}^{\mathrm{H}}\mathbf{v}) \\
&= \left(\|\mathbf{x}\|_2^2 + \|\mathbf{v}\|_2^2\right)^2 - 4\left[\Re^2(\mathbf{x}^{\mathrm{H}}\mathbf{v}) + \Im^2(\mathbf{x}^{\mathrm{H}}\mathbf{v})\right] \\
&= \left(\|\mathbf{x}\|_{22}^{22} + \|\mathbf{v}\|_2^2\right)^2 - 4|\mathbf{x}^{\mathrm{H}}\mathbf{v}|^2 \\
&= \left(\|\mathbf{x}\|_2^2 + \|\mathbf{v}\|_2^2 + 2|\mathbf{x}^{\mathrm{H}}\mathbf{v}|\right)\left(\|\mathbf{x}\|_2^2 + \|\mathbf{v}\|_2^2 - 2|\mathbf{x}^{\mathrm{H}}\mathbf{v}|\right) \\
&\geq \|\mathbf{v}\|_2^2\left(\|\mathbf{x}\|_2^2 + \|\mathbf{v}\|_2^2 - 2|\mathbf{x}^{\mathrm{H}}\mathbf{v}|\right) \\
&= \|\mathbf{v}\|_2^2\mathrm{dist}^2(\mathbf{x}, \mathbf{v}).
\end{aligned}$$

Algorithm 3.2 Stochastic Prox-linear SE Solver

1: **Input** data $\{(z_m, \mathbf{H}_m)\}_{m=1}^{M}$, step size $\mu > 0$, initialization $\mathbf{v}_0 \in \mathbb{C}^N$, solution accuracy $\epsilon > 0$, and set $t = 0$

2: **Prepare** the power flow data $\{(z_m, \mathbf{H}_m)\}_{m=N+1}^{M}$ according to (3.1), (3.23–3.24), (3.26–3.27), (3.30–3.31), and (3.32–3.33) to obtain $\{(\breve{z}_m, \mathbf{h}_m)\}_{m=N+1}^{M}$ based on $\{z_m = |v_m|^2\}_{m=1}^{N}$

3: **Repeat**

4: **Draw** $m_t \in \{1, 2, \ldots, M\}$ uniformly at random

5: **Evaluate** Evaluate $\mathbf{a}_{m_t,t}$ and $b_{m_t,t}$ in (3.40a) and (3.40b)

6: **Update** $\mathbf{x}_{t+1}$ via (3.45)

7: $t = t + 1$

8: **Until**

9: **Return** $\mathbf{x}_t$

Taking the square root from both sides of the inequality yields the statement of Lemma 3.1

Upon defining $\mathbf{w} := \mathbf{x} - \mathbf{x}_t$, $\mathbf{a} := \mathbf{a}_{m_t,t}$, and $b := b_{m_t,t} - 2\Re(\mathbf{a}_{m_t,t}^{\mathrm{H}}\mathbf{x}_t)$ in Proposition 3.2, one can readily find the solution to (3.41) as follows

$$\mathbf{x}_{t+1} = \mathbf{x}_t + \mathrm{proj}_{\mu_t}\left(\frac{b_{m_t,t} - 2\Re\left(\mathbf{a}_{m_t,t}^{\mathrm{H}}\mathbf{x}_t\right)}{\left\|\mathbf{a}_{m_t,t}\right\|_2^2}\right)\mathbf{a}_{m_t,t}.$$

Substituting $\mathbf{a}_{m_t}$ and b_{m_t} of (3.40a) and (3.40b) into the last equality leads to our stochastic prox-linear SE solver

$$\mathbf{x}_{t+1} = \mathbf{x}_t + \mathrm{proj}_{\mu_t}\left(\frac{\breve{z}_{m_t} - \left|\mathbf{h}_{m_t}^{\mathrm{H}}\mathbf{x}_t\right|^2}{4\left|\mathbf{h}_{m_t}^{\mathrm{H}}\mathbf{x}_t\right|^2 \cdot \left\|\mathbf{h}_{m_t}\right\|_2^2}\right) \cdot 2\left(\mathbf{h}_{m_t}^{\mathrm{H}}\mathbf{x}_t\right)\mathbf{h}_{m_t}. \tag{3.45}$$

We present our second (stochastic) prox-linear SE solver in Algorithm 3.2 for future reference. To provide a clear understanding of the computational complexity, we will detail the exact number of complex scalar operations (e.g., additions, multiplications) required for each stochastic prox-linear SE iteration of (3.45) below. Depending on whether $\mathbf{h}_{m_t}$ has 1 or 2 nonzero entries, the following statements hold true. If $\mathbf{h}_{m_t}$ has 1 (2) nonzero entries, evaluating $\left|\mathbf{h}_{m_t}^{\mathrm{H}}\mathbf{x}_t\right|^2$ requires 2 (4) operations, while computing $\left\|\mathbf{h}_{m_t}\right\|_2^2$ requires 1 (3) operations. Additionally, there are another 5 operations for the remaining calculations, summing up to a total of 8 (12) operations. In essence, per iteration of Algorithm 3.2 (see Steps 4–6) necessitates only approximately 12 complex scalar operations. Remarkably, this per-iteration complexity of $O(1)$ remains constant regardless of the size of the power network under consideration, or more precisely, the system size N. It is evident that this $O(1)$ per-iteration complexity is well-suited for both large and massive-size power networks.

3.4 Convergence Analysis and Exact Recovery

In this section, we initiate our exploration by presenting convergence assurances for the proposed prox-linear SE solvers. To maintain clarity, our primary attention will be directed towards the deterministic prox-linear algorithm 3.1, although it is worth noting that convergence can also be established for Algorithm 3.2 in a probabilistic sense. For a more in-depth examination of the latter, we refer interested readers to (Duchi and Ruan 2018). It is important to mention that, subject to specific conditions on the loss function ℓ, we are able to establish exact recovery results as well.

Recall our loss function $c(\mathbf{s}(\mathbf{x})) = \frac{1}{M}\left\|\breve{\mathbf{z}} - |\mathbf{H}\mathbf{x}|^2\right\|_1$, where $c(\mathbf{u}) = \|\mathbf{u}\|_1 : \mathbb{R}^M \to \mathbb{R}$, and $\mathbf{s}(\mathbf{x}) = \frac{1}{M}(\breve{\mathbf{z}} - |\mathbf{H}\mathbf{x}|^2) : \mathbb{C}^N \to \mathbb{R}^M$. It is easy to verify for any $\mathbf{u}, \mathbf{v} \in \mathbb{R}^N$ that it holds

$$|c(\mathbf{u}) - c(\mathbf{v})| \leq \sum_{n=1}^{M} |u_m - v_m| = \|\mathbf{u} - \mathbf{v}\|_1 \leq \sqrt{M}\,\|\mathbf{u} - \mathbf{v}\|_2,$$

where the last inequality arises from the equivalence of norms. By definition, this asserts that c is $\sqrt{M}$-Lipschitz continuous. Focusing now on the complex Jacobian $\nabla_{\mathrm{x}}\mathbf{s}$ for any $\mathbf{x}$ and $\mathbf{y} \in \mathbb{C}^N$, we deduce

$$\|\nabla_{\mathrm{x}}\mathbf{s}(\mathbf{x}) - \nabla_{\mathrm{x}}\mathbf{s}(\mathbf{y})\|_2 = \frac{1}{M}\left\|\mathbf{H}^{\mathrm{H}}\mathbf{H}(\mathbf{x} - \mathbf{y})\right\|_2 \leq L\|\mathbf{x} - \mathbf{y}\|_2,$$

with $L := (2/M)\lambda_{\max}(\mathbf{H}^{\mathrm{H}}\mathbf{H})$, which confirms that $\nabla_{\mathrm{x}}\mathbf{s}$ is L-Lipschitz continuous.

Referring to the findings presented in (Lewis and Wright 2016) it is evident that our deterministic prox-linear SE solver, when equipped with a constant step size $\mu \leq 1/(L\sqrt{M}) = \sqrt{M}/(2\lambda_{\max}(\mathbf{H}^{\mathrm{H}}\mathbf{H}))$, converges towards a stationary point of the objective function $\ell(\mathbf{x})$ as defined in (3.35).

In the subsequent sections, we will outline specific conditions that must be satisfied by the function ℓ in order to ensure the precise recovery of power system states through the utilization of our prox-linear SE solvers. Unlike the constraints imposed in (Duchi and Ruan 2019), which exclusively pertain to optimization over real-valued variables, we will introduce two complementary conditions that apply to both $\ell(\mathbf{x})$ and its linearization $\ell_{\mathbf{x}}(\mathbf{y})$, accommodating complex-valued variables $\mathbf{x}, \mathbf{y} \in \mathbb{C}^N$.

If ℓ is defined over real-valued vectors, namely, $\ell(\mathbf{x}) : \mathbb{R}^N \to \mathbb{R}$, we can readily have the following stability condition for establishing fast convergence: for any given $\mathbf{v} \in \mathbb{R}^N$, there exists a constant $\rho > 0$ such that $\ell(\mathbf{x}) - \ell(\mathbf{v}) \geq \rho\|\mathbf{x} - \mathbf{v}\|_2\|\mathbf{x} + \mathbf{v}\|_2$ holds for all $\mathbf{x} \in \mathbb{R}^N$. However, this condition, while essential for ensuring swift convergence, is not readily applicable to functions involving complex-valued variables. In the complex domain, it is evident that if $\mathbf{v} \in \mathbb{C}^N$ is an optimal solution to (3.35), then any complex exponential of it, denoted as $e^{j\phi}\mathbf{v}$, where ϕ ranges from 0 to 2π, is also an optimal solution. This inherent ambiguity leads us to define the Euclidean distance between any estimate $\mathbf{x} \in \mathbb{C}^N$ and the true state vector $\mathbf{v} \in \mathbb{C}^N$

as $\text{dist}(\mathbf{x}, \mathbf{v}) := \min_{\phi \in [0,2\pi)} \left\| \mathbf{x} - e^{j\phi}\mathbf{v} \right\|_2$. This definition can be likened to imposing a zero phase angle at the reference bus, a standard practice in power flow problems to eliminate phase ambiguity. By drawing upon a result established for stable phase retrieval in Balan, 2016, we formulate the following stability condition tailored for functions $\ell(\mathbf{x}) : \mathbb{C}^N \to \mathbb{R}$. This condition remains applicable even in practical cases where measurements may be affected by noise or adversarial corruption.

Condition 2 There exists some constant $\rho > 0$ such that the inequality holds for all $\mathbf{x} \in \mathbb{C}^N$

$$\ell(\mathbf{x}) - \ell(\mathbf{v}) \geq \rho \sqrt{\|\mathbf{x} - \mathbf{v}\|_2^2 \|\mathbf{x} + \mathbf{v}\|_2^2 - 4\left|\Im(\mathbf{x}^{\mathrm{H}}\mathbf{v})\right|^2}. \tag{3.46}$$

Evidently, in the noiseless case where the measurements $\breve{\mathbf{z}} = |\mathbf{H}\mathbf{v}|^2$, the loss function $\ell(\mathbf{v})$ evaluates to zero. Similar to the real-valued case studied in (Duchi and Ruan 2019), our Condition 2 is instrumental in establishing fast convergence of the prox-linear algorithm for optimizing functions of complex-valued variables. In addition to Condition 2, we require a condition pertaining to the linearization $\ell_{\mathbf{x}}(\mathbf{y}) : \mathbb{C}^N \to \mathbb{R}$ of $\ell(\mathbf{x}) = c(\mathbf{s}(\mathbf{x}))$ around $\mathbf{x}$, defined as follows

$$\ell_{\mathbf{x}}(\mathbf{y}) := c\big(\mathbf{s}(\mathbf{x}) + 2\Re\big(\nabla_{\mathbf{x}}^{\mathrm{H}}\mathbf{s}(\mathbf{x})(\mathbf{y} - \mathbf{x})\big)\big). \tag{3.47}$$

Condition 3 There exists a constant $L < +\infty$ such that the inequality holds for all $\mathbf{x} \in \mathbb{C}^N$

$$|\ell(\mathbf{y}) - \ell_{\mathbf{x}}(\mathbf{y})| \leq \frac{L}{2}\|\mathbf{x} - \mathbf{y}\|_2^2. \tag{3.48}$$

This condition essentially ensures that the locally linearized convex approximation $\ell_{\mathbf{x}}(\mathbf{y})$ is closely quadratic [cf. (3.48)] to the non-convex function $\ell(\mathbf{y})$. Notably, the ℓ_1-based PSSE cost function ℓ in (3.35) inherently satisfies Condition 3 on a global scale. To demonstrate this, let's initially express $\ell_{\mathbf{x}}(\mathbf{y})$ based on the definition in (3.47)

$$\ell_{\mathbf{x}}(\mathbf{y}) = \frac{1}{M}\sum_{m=1}^{M}\left| |\mathbf{h}_m^{\mathrm{H}}\mathbf{x}|^2 - \breve{z}_m + 2\Re\big(\mathbf{x}^{\mathrm{H}}\mathbf{h}_m\mathbf{h}_m^{\mathrm{H}}(\mathbf{y} - \mathbf{x})\big)\right|.$$

On the other hand, for any $m \in \{1, 2, \ldots, M\}$ and $\mathbf{y} \in \mathbb{C}^N$, the following holds true

$$\left|\mathbf{h}_m^{\mathrm{H}}\mathbf{y}\right|^2 = \left|\mathbf{h}_m^{\mathrm{H}}\mathbf{x}\right|^2 + 2\Re\big(\mathbf{x}^{\mathrm{H}}\mathbf{h}_m\mathbf{h}_m^{\mathrm{H}}(\mathbf{y} - \mathbf{x})\big) + \left|\mathbf{h}_m^{\mathrm{H}}(\mathbf{y} - \mathbf{x})\right|^2.$$

Subtracting $\breve{z}_m$ from both sides, summing from $m = 1$ to M, and leveraging the triangle inequality, we have that

$$\ell(\mathbf{y}) = \frac{1}{M}\sum_{m=1}^{M}\left||\mathbf{h}_m^{\mathrm{H}}\mathbf{y}|^2 - \breve{z}_m\right| \leq \ell_{\mathbf{x}}(\mathbf{y}) + \frac{1}{M}\sum_{m=1}^{M}\left|\mathbf{h}_m^{\mathrm{H}}(\mathbf{y}-\mathbf{x})\right|^2,$$

$$\ell(\mathbf{y}) = \frac{1}{M}\sum_{m=1}^{M}\left||\mathbf{h}_m^{\mathrm{H}}\mathbf{y}|^2 - \breve{z}_m\right| \geq \ell_{\mathbf{x}}(\mathbf{y}) - \frac{1}{M}\sum_{m=1}^{M}\left|\mathbf{h}_m^{\mathrm{H}}(\mathbf{y}-\mathbf{x})\right|^2.$$

Rewriting the last terms in matrix–vector form yields

$$|\ell(\mathbf{y}) - \ell_{\mathbf{x}}(\mathbf{y})| \leq (\mathbf{y}-\mathbf{x})^{\mathrm{H}}(\frac{1}{M}\mathbf{H}^{\mathrm{H}}\mathbf{H})(\mathbf{y}-\mathbf{x}) \leq \frac{L}{2}\|\mathbf{y}-\mathbf{x}\|_2^2, \tag{3.49}$$

which proves that Condition 3 is satisfied globally by the ℓ_1 loss in (3.35). Furthermore, it is possible to choose $L = \lambda_{\max}(\mathbf{H}^{\mathrm{H}}\mathbf{H}/M)$, which corresponds to the largest eigenvalue of the matrix $\mathbf{H}^{\mathrm{H}}\mathbf{H}/M$.

Under the stipulated Conditions 2 and 3, we can now delve into the exact recovery guarantees for the deterministic prox-linear SE solver outlined in Algorithm 3.1. The forthcoming result implies that our proposed prox-linear SE solver, as delineated in Algorithm 3.1, can achieve exact recovery of power system states at a quadratic rate, provided the specified conditions are met.

Theorem 3.1 *Let Conditions 2 and 3 hold. Assuming that the quadratic program* (3.38) *is solved exactly per iteration of Algorithm 3.1, the successive prox-linear SE iterates* $\mathbf{x}_t$ *satisfy*

$$\frac{\mathrm{dist}(\mathbf{x}_t, \mathbf{v})}{\|\mathbf{v}\|_2} \leq \frac{\rho}{L}\left(\frac{L}{\rho}\cdot\frac{\mathrm{dist}(\mathbf{x}_0, \mathbf{v})}{\|\mathbf{v}\|_2}\right)^{2^t}. \tag{3.50}$$

Proof Notice that the regularized function $g(\mathbf{x}) := \ell_{\mathbf{x}_t}(\mathbf{x}) + \frac{L}{2}\|\mathbf{x}-\mathbf{x}_t\|_2^2$ exhibits L-strong convexity in $\mathbf{x} \in \mathbb{C}^N$. Its minimum coincides with $\mathbf{x}_{t+1}$ as defined in (3.38). Employing the standard optimality conditions for strongly convex minimization, it follows that $g(\mathbf{x}_{t+1}) \leq g(\mathrm{e}^{j\angle\mathbf{x}_t^{\mathrm{H}}\mathbf{v}}\mathbf{v}) - \frac{L}{2}\left\|\mathrm{e}^{j\angle\mathbf{x}_t^{\mathrm{H}}\mathbf{v}}\mathbf{v} - \mathbf{x}_{t+1}\right\|_2^2$, that is

$$\begin{aligned}&\ell_{\mathbf{x}_t}(\mathbf{x}_{t+1}) + \frac{L}{2}\|\mathbf{x}_{t+1}-\mathbf{x}_t\|_2^2\\ &\leq \ell_{\mathbf{x}_t}(\mathrm{e}^{j\angle\mathbf{x}_t^{\mathrm{H}}\mathbf{v}}\mathbf{v}) + \frac{L}{2}\left\|\mathrm{e}^{j\angle\mathbf{x}_t^{\mathrm{H}}\mathbf{v}}\mathbf{v} - \mathbf{x}_t\right\|_2^2 - \frac{L}{2}\left\|\mathrm{e}^{j\angle\mathbf{x}_t^{\mathrm{H}}\mathbf{v}}\mathbf{v} - \mathbf{x}_{t+1}\right\|_2^2\end{aligned} \tag{3.51}$$

Recalling now Condition 3, we have that

$$\ell(\mathbf{x}_{t+1}) \leq \ell_{\mathbf{x}_t}(\mathbf{x}_{t+1}) + \frac{L}{2}\|\mathbf{x}_{t+1} - \mathbf{x}\|_{t2}^2, \tag{3.52a}$$

$$\ell_{\mathbf{x}_t}(\mathrm{e}^{j\angle\mathbf{x}_t^{\mathrm{H}}\mathbf{v}}\mathbf{v}) \leq \ell(\mathrm{e}^{j\angle\mathbf{x}_t^{\mathrm{H}}\mathbf{v}}\mathbf{v}) + \frac{L}{2}\left\|\mathrm{e}^{j\angle\mathbf{x}_t^{\mathrm{H}}\mathbf{v}}\mathbf{v} - \mathbf{x}_t\right\|_2^2. \tag{3.52b}$$

Substituting (3.52a) and (3.52b) into (3.51) gives rise to

$$\begin{aligned}\ell(\mathbf{x}_{t+1}) &\leq \ell_{\mathbf{x}_t}(\mathrm{e}^{j\angle \mathbf{x}_t^{\mathrm{H}}\mathbf{v}}\mathbf{v}) + \frac{L}{2}\left\|\mathrm{e}^{j\angle \mathbf{x}_t^{\mathrm{H}}\mathbf{v}}\mathbf{v} - \mathbf{x}_t\right\|_2^2 - \frac{L}{2}\left\|\mathrm{e}^{j\angle \mathbf{x}_t^{\mathrm{H}}\mathbf{v}}\mathbf{v} - \mathbf{x}_{t+1}\right\|_2^2\\ &\leq \ell(\mathrm{e}^{j\angle \mathbf{x}_t^{\mathrm{H}}\mathbf{v}}\mathbf{v}) + L\left\|\mathrm{e}^{j\angle \mathbf{x}_t^{\mathrm{H}}\mathbf{v}}\mathbf{v} - \mathbf{x}_t\right\|_2^2 - \frac{L}{2}\left\|\mathrm{e}^{j\angle \mathbf{x}_t^{\mathrm{H}}\mathbf{v}}\mathbf{v} - \mathbf{x}_{t+1}\right\|_2^2,\end{aligned}$$

which, in conjunction with $\ell(\mathrm{e}^{j\angle \mathbf{x}_t^{\mathrm{H}}\mathbf{v}}\mathbf{v}) = \ell(\mathbf{v})$ in (3.35), yields

$$L\left\|\mathrm{e}^{j\angle \mathbf{x}_t^{\mathrm{H}}\mathbf{v}}\mathbf{v} - \mathbf{x}_t\right\|_2^2 \geq \ell(\mathbf{x}_{t+1}) - \ell(\mathbf{v}) + \frac{L}{2}\left\|\mathrm{e}^{j\angle \mathbf{x}_t^{\mathrm{H}}\mathbf{v}}\mathbf{v} - \mathbf{x}_{t+1}\right\|_2^2. \tag{3.53}$$

Invoking further Condition 2 in (3.53), we have that

$$\begin{aligned}&L\left\|\mathrm{e}^{j\angle \mathbf{x}_t^{\mathrm{H}}\mathbf{v}}\mathbf{v} - \mathbf{x}_t\right\|_2^2\\ &\geq \frac{L}{2}\left\|\mathrm{e}^{j\angle \mathbf{x}_t^{\mathrm{H}}\mathbf{v}}\mathbf{v} - \mathbf{x}_{t+1}\right\|_2^2 \quad + \rho\sqrt{\|\mathbf{x}_{t+1} - \mathbf{v}\|_2^2\|\mathbf{x}_{t+1} + \mathbf{v}\|_2^2 - 4\left|\Im(\mathbf{x}_{t+1}^{\mathrm{H}}\mathbf{v})\right|^2},\end{aligned} \tag{3.54}$$

where the last term can be replaced with its bound $\rho\|\mathbf{v}\|_2\mathrm{dist}(\mathbf{x}_{t+1}, \mathbf{v})$ established in Lemma 3.1. Upon dropping the non-negative term $\frac{L}{2}\left\|\mathrm{e}^{j\angle \mathbf{x}_t^{\mathrm{H}}\mathbf{v}}\mathbf{v} - \mathbf{x}_{t+1}\right\|_2^2$, and recalling the definition of $\mathrm{dist}(\mathbf{x}_t, \mathbf{v})$, we immediately have

$$\rho\|\mathbf{v}\|_2\mathrm{dist}(\mathbf{x}_{t+1}, \mathbf{v}) \leq L\mathrm{dist}^2(\mathbf{x}_t, \mathbf{v}),$$

dividing both sides of which by $\rho\|\mathbf{v}\|_2^2$ yields

$$\frac{\mathrm{dist}(\mathbf{x}_{t+1}, \mathbf{v})}{\|\mathbf{v}\|_2} \leq \frac{L}{\rho}\cdot\frac{\mathrm{dist}^2(\mathbf{x}_t, \mathbf{v})}{\|\mathbf{v}\|_2^2} = \frac{\rho}{L}\left(\frac{L}{\rho}\cdot\frac{\mathrm{dist}(\mathbf{x}_t, \mathbf{v})}{\|\mathbf{v}\|_2}\right)^2.$$

Applying the above inequality successively from the initialization $\mathbf{x}_0$ for t iterations through $\mathbf{x}_t$ gives rise to (3.50), concluding the proof.

Remark 3.1 (*Exact recovery*) If the initialization $\mathbf{x}_0$ for the iterations meets the accuracy condition $\mathrm{dist}(\mathbf{x}_0, \mathbf{v}) < (\rho/L)\|\mathbf{v}\|_2$, the prox-linear SE solver achieves an exact recovery of the true state vector $\mathbf{v} \in \mathbb{C}^N$. In terms of initialization strategies, several approaches are suitable, and we discuss three of them below. Given that power systems typically operate close to the flat voltage profile $\mathbf{1}$, it is reasonable to initialize the algorithm with $\mathbf{x}_0 = \mathbf{1}$. Additionally, as the voltage magnitudes at all buses are assumed to be available, one can employ the voltage magnitude vector $\mathbf{x}_0 = |\mathbf{v}|$ as the initializer. Alternatively, it is also feasible to initialize with the estimate obtained by solving the linearized DC power flow equations.

Remark 3.2 (*Quadratic convergence rate*) When $\text{dist}(\mathbf{x}_0, \mathbf{v}) < (\rho/L)\|\mathbf{v}\|_2$ holds, our prox-linear SE algorithm exhibits quadratic convergence to the globally optimal solution of the non-convex and non-smooth optimization problem (3.35).

In other words, to achieve a solution $\mathbf{x}_t$ with an (at most) ϵ-relative error, denoted as $\text{dist}(\mathbf{x}_t, \mathbf{v})/\|\mathbf{v}\|_2 \leq \epsilon$, we need to execute Algorithm 3.1 for approximately $\log_2 \log_2(1/\epsilon)$ iterations. This is equivalent to solving around $\log_2 \log_2(1/\epsilon)$ convex quadratic programs, as seen in (3.38). In practical terms, this translates to solving approximately $5 \sim 10$ such quadratic programs.

Remark 3.3 Under the condition $\text{dist}(\mathbf{x}_0, \mathbf{v}) < (\rho/L)\|\mathbf{v}\|_2$, it is worth noting that Condition 2 can be replaced by a condition that only requires the function $\ell(\mathbf{x})$ to locally satisfy the inequality (3.44) for all $\mathbf{x}$ within a neighborhood of $\mathbf{v}$ defined by $\text{dist}(\mathbf{x}, \mathbf{v}) < (\rho/L)\|\mathbf{v}\|_2$.

3.5 Numerical Tests

In this section, we conduct a series of numerical tests to assess the performance of our approach and compare it with the widely used LS-based Gauss–Newton method, often considered the 'workhorse' in PSSE. We performed simulations on various power network benchmarks, specifically, the IEEE 14-, 118-, and 300-bus systems, using the MATPOWER toolbox in MATLAB (Christie 2000; Zimmerman et al. 2010). The Gauss–Newton iterations were implemented by using the embedded SE function 'doSE.m' in MATPOWER. To carefully isolate the relative performance of the iterative algorithms, rather than initialization employed, all simulated schemes were initialized with the flat voltage vector (i.e., the all-one vector) in all reported experiments.

3.5.1 Tests with Zero Noise

Our initial experiment focuses on assessing the performance of Algorithm 3.1 in terms of exact recovery and convergence when applied to noiseless data from the IEEE 14-, 118-, and 300-bus test systems. For this experiment, we introduced variability into the actual voltage magnitudes (expressed in per unit) and angles (measured in radians) at each bus. Specifically, these parameters were uniformly distributed over the intervals $[0.9, 1.1]$ and $[-0.1\pi, 0.1\pi]$, respectively. The voltage magnitude squares at all buses as well as the active power flows across all lines were measured. The quadratic programs in Step 5 of Algorithm 3.1 were solved by the standard convex programming solver (SeDuMi Sturm 1999) with a constant step size of $\mu_t = 1,000$. Algorithm 3.1 terminates either when a maximum number 20 of iterations are simulated, or when the normalized distance between two consecutive iterates becomes smaller than 10^{-10}, namely $\text{dist}(\mathbf{x}_t, \mathbf{x}_{t-1})/\sqrt{N} \leq 10^{-10}$. A total of 100 Monte Carlo (MC) runs were carried out. Figures 3.1, 3.2 and 3.3 plot the normalized estimation errors $\text{dist}(\mathbf{x}_t, \mathbf{v})/\sqrt{N}$ for the 100 MC realizations on the simulated three systems, whose corresponding L values are 0.9980, 3.0201, and 6.3102. Further, Fig. 3.4 depicts the convergence of the normalized estimation error of Algorithm 3.1 for the 100 runs on the 14-bus system.} Evidently, Algorithm 3.1 achieves exact recovery over the 100 runs, and enjoys quadratic convergence in this noiseless setting, validating our theoretical findings in Theorem 3.1.

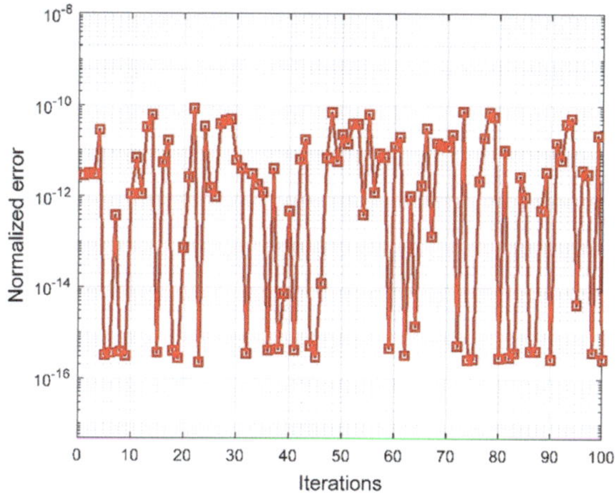

Fig. 3.1 Exact recovery performance of Algorithm 3.1 for the IEEE 14-bus system (noiseless case)

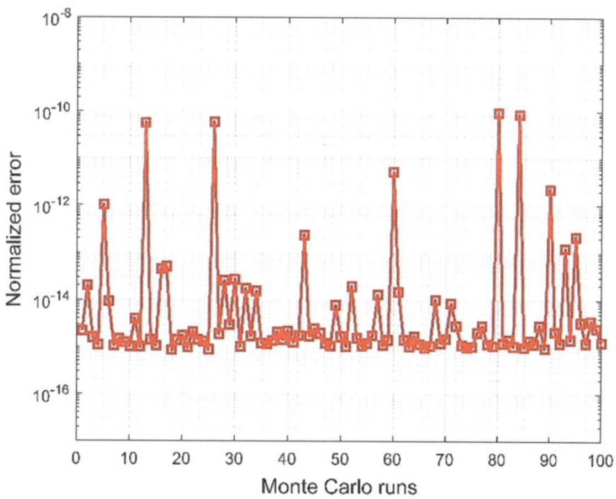

Fig. 3.2 Exact recovery performance of Algorithm 3.1 for the IEEE 118-bus system (noiseless case)

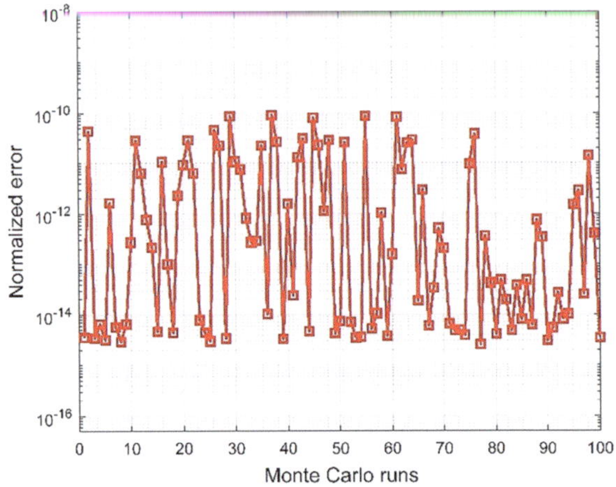

Fig. 3.3 Exact recovery performance of Algorithm 3.1 for the IEEE 300-bus system (noiseless case)

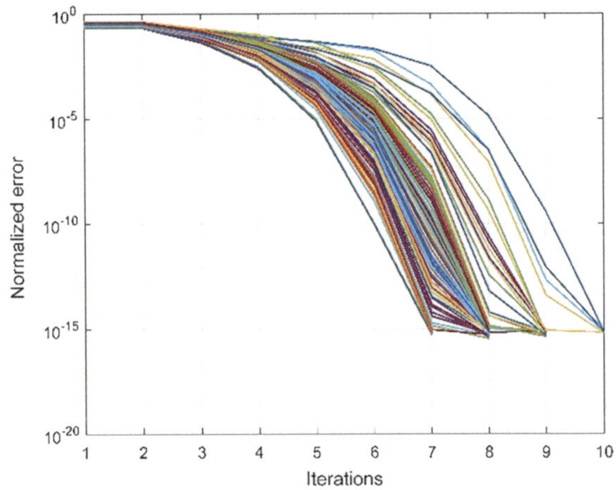

Fig. 3.4 Quadratic convergence of Algorithm 3.1 in 100 runs for the IEEE 14-bus system (noiseless case)

3.5.2 Tests With Outlying Measurements

An inherent advantage attributed to the ℓ_1-based loss function presented in Eq. (3.35) is its resilience in the face of outliers. In the following assessment, we delve into Algorithm 3.1's robustness concerning exact recovery in cases involving measurements tainted by outliers. To conduct this evaluation, we utilized the IEEE 118-bus system, maintaining its default voltage profile intact. We considered measurements of voltage magnitudes at all buses and the active and reactive power flows across all lines. Given the common practice of maintaining nodal voltage magnitudes close to unity in transmission networks, we assumed that only the power meters might be susceptible to compromise. Our simulation involved running 100 MC iterations, each introducing an outlier by deliberately manipulating and amplifying the measurement

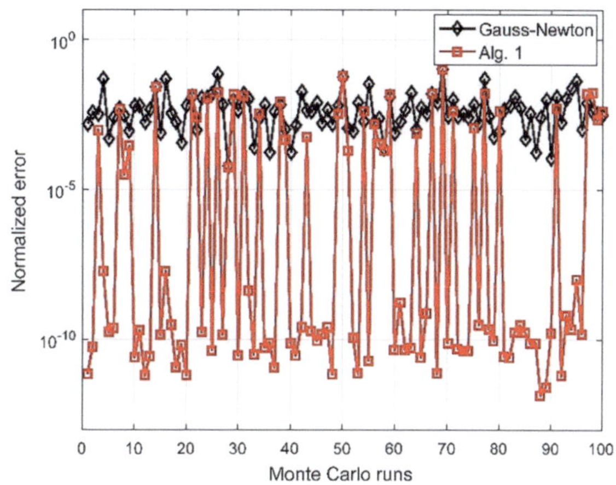

Fig. 3.5 Recovery performance for the IEEE 118-bus system with 1 outlying measurement (noiseless case)

of one randomly selected power flow meter to five times its original value. The objective was to assess Algorithm 3.1's resilience and its ability to achieve exact recovery under these challenging outlier conditions.

Both the Gauss–Newton and Algorithm 3.1 were simulated in this experiment, whose corresponding normalized estimation errors for the 100 MC realizations are presented in Fig. 3.5. It is self-evident from the plots that the Gauss–Newton method is not robust to outlying measurements, whereas our proposed prox-linear scheme in Algorithm 3.1 can identify and automatically reject the bad data, yielding exact recovery of the true states in most cases even under adversarial attacks.

3.5.3 *Tests with Additive Noise and Outliers*

In the third experiment, we conducted an evaluation of Algorithm 3.1's robust estimation performance in comparison to the Gauss–Newton method. This assessment was carried out in a setting characterized by the simultaneous presence of both additive noise and outliers. We opted for the IEEE 300-bus benchmark system, employing its default voltage profile, as the simulation framework. All measurements, encompassing active and reactive power flows in addition to voltage magnitudes, were considered. The additive noise component was introduced through the generation of independent normal distributions, featuring zero-mean characteristics and standard deviations of 0.004 and 0.008 for the voltage magnitude and line flow measurements, respectively, in accordance with previous research (Abur and Celik 1991). In conjunction with the additive noise, a further element was introduced into the simulation: five percent of all measured power flows were deliberately corrupted in a random and uniform manner. These 'outliers' were drawn independently from a

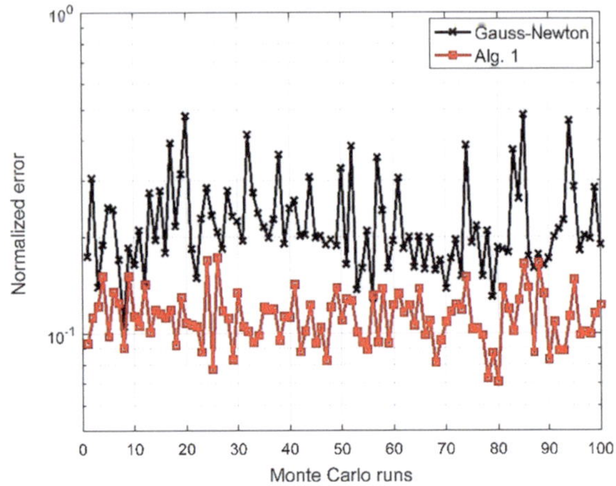

Fig. 3.6 Estimation performance for the IEEE 300-bus system under additive noise and 5% outlying measurements

Gaussian distribution characterized by a zero-mean and a standard deviation of 5. The figures reported in Fig. 3.6 showcase the normalized estimation errors resulting from 100 independent MC realizations, comparing the performance of the Gauss–Newton method and Algorithm 3.1. It is abundantly clear from the data that our developed algorithm consistently demonstrates superior robustness in the face of both additive noise and outlying measurements, surpassing the LS-based Gauss–Newton approach.

Chapter 4
State Estimation and Forecasting via Deep Unrolled Neural Networks

The increasing integration of distributed renewable generators, electric vehicles, and demand response initiatives has introduced significant challenges to modern power grids. These challenges manifest in the form of unparalleled load peaks and voltage fluctuations. With knowledge of grid parameters and a collection of measurements sourced from the SCADA system, the objective of PSSE is to ascertain the unobserved system state, specifically the complex voltages at all network buses (Wood et al. 2013; Kekatos et al. 2017). Nonetheless, the existing optimization-based PSSE methodologies often demand a substantial number of iterations and impose a significant computational burden. Moreover, they face increasing challenges posed by the expanding dynamics and scale of modern power systems. These factors underscore the need for innovative approaches to enable real-time, large-scale PSSE. To address this need, the utilization of plain feed-forward neural networks (FNNs) for PSSE has been explored in previous works (Barbeiro et al. 2014; Mestav et al. 2018; Zamzam et al. 2019a; Zamzam and Sidiropoulos 2020). These FNNs can be trained offline using historical data or simulated samples and subsequently deployed for real-time PSSE, as the inference process involves just a few matrix–vector multiplications. Power system state forecasting has traditionally been explored using methods such as (extended) Kalman filtering and moving horizon approaches, as documented in references (Debs and Larson 1970; Da Silva et al. 1983; Hassanzadeh and Evrenosoğlu 2012; Kekatos et al. 2017). Additionally, first-order vector auto-regressive (VAR) modeling has been employed in this context, as outlined in reference (Hassanzadeh et al. 2015). Unfortunately, the growth of the number of FNN parameters exhibits a linear relationship with the length of input sequences. This characteristic discourages FNNs from effectively capturing long-term dependencies within voltage time series.

To overcome these obstacles, this chapter proposes the utilization of physics-inspired DNNs for real-time power system monitoring. We have developed a novel, model-specific DNNs that leverages an unrolled iterative solver originally designed using an accurate AC model for real-time PSSE, requiring only offline training and

G. Wang et al., *Data-driven Optimization and Control for Autonomous Energy Systems*,
https://doi.org/10.1007/978-981-95-1782-4_4

minimal adjustments. To proactively sense the system state and enhance the robustness of the DNNs estimator, we have introduced deep recurrent neural networks (RNNs) for power system state prediction. These deep RNNs exploit long-term nonlinear dependencies present in historical voltage time series for efficient prediction and ease of implementation. Numerical test results demonstrate significant performance improvements of the DNNs-based estimation and prediction methods compared to existing alternatives. In experiments conducted using real load data from the IEEE 118-bus benchmark system, the performance of the novel model-specific PSSE approach based on DNNs was nearly an order of magnitude better than competing solutions, including widely adopted Gauss–Newton PSSE solvers.

4.1 System Modeling and Problem Formulation

Consider a power network consisting of N buses that can be modeled as a graph $G := \{N, L\}$, where $N := \{1, 2, \ldots, N\}$ comprises all buses, and $L := \{(n, n')\} \in N \times N$ collects all lines; see Fig. 4.1. For each bus $n \in N$, let $V_n := v_n^r + jv_n^i$ denote its corresponding complex voltage with magnitude $|V_n|$, and P_n (Q_n) denote the reactive power injection. For each line $(n, n') \in L$, let $P_{nn'}^f$ ($Q_{nn'}^f$) denote the (re)active power flow seen at the 'forwarding' end, and $P_{nn'}^t$ ($Q_{nn'}^t$) denote the (re)active power flow at the 'terminal' end.

To perform PSSE, suppose M_t system variables are measured at time instant t. For a compact representation, let $z_t := [\{|V_{n,t}|^2\}_{n\in N_t^o}, \{P_{n,t}\}_{n\in N_t^o}, \{Q_{n,t}\}_{n\in N_t^o}, \{P_{nn',t}^f\}_{(n,n')\in E_t^o}, \{Q_{nn',t}^t\}_{(n,n')\in E_t^o}]^{\mathrm{T}}\{Q_{nn',t}^f\}_{(n,n')\in E_t^o}, \{P_{nn',t}^t\}_{(n,n')\in E_t^o}$, be the measurement vector that collects all measured quantities at time t, where N_t^o and E_t^o denote the locations where nodal and line quantities are measured, respectively.

Per time instant t, PSSE aims to recover the system state vector $v_t := [v_{1,t}^r\, v_{1,t}^i \cdots v_{N,t}^r\, v_{N,t}^i]^{\mathrm{T}} \in \mathbb{R}^{2N}$ (in rectangular coordinates) from generally noisy measurement z_t. For brevity, the subscript t of z_t and v_t will be omitted when discussing PSSE in Sects. 4.1 and 4.2. Mathematically, the PSSE task can be posed as follows.

Given measurements $\mathbf{z} := \{z_m\}_{m=1}^M$ and $\{\mathbf{H}_m \in \mathbb{R}^{2N\times 2N}\}_{m=1}^M$ obeying the following physical model

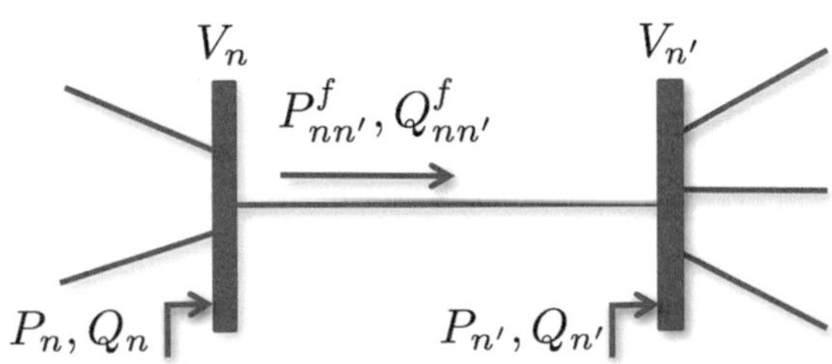

Fig. 4.1 Bus n is connected to bus n' via line (n, n')

$$z_m = \mathbf{v}^{\mathrm{T}}\mathbf{H}_m\mathbf{v} + \epsilon_m, \qquad \forall m = 1, \ldots, M, \tag{4.1}$$

our objective is the recovery of $\mathbf{v} \in \mathbb{R}^{2N}$, where $\{\epsilon_m\}_{m=1}^{M}$ encompass the effects of measurement noise and modeling inaccuracies.

Adopting the LAV error criterion that is known to be robust to outliers, the ensuing LAV estimate is sought (Wood et al. 2013)

$$\hat{v} := \arg \min_{\mathbf{v}\in\mathbb{R}^{2N}} \frac{1}{M} \sum_{m=1}^{M} \left| z_m - \mathbf{v}^{\mathrm{T}}\mathbf{H}_m\mathbf{v} \right|, \tag{4.2}$$

for which various solvers have been developed (Abur and Celik 1991; Jabr and Pal 2004; Kekatos et al. 2017). Notably, a recent prox-linear solver, introduced in (Wang et al. 2019b), has garnered significant attention due to its well documented merits. These advantages include provably fast (locally quadratic) convergence and efficiency in handling the non-smoothness and non-convexity inherent in (4.2). To elaborate, the prox-linear solver commences with an initial vector $\mathbf{v}_0$ and subsequently engages in iterative steps to minimize the regularized and 'locally linearized' (relaxed) cost described in (4.2). This iterative process is further detailed in (Burke and Ferris 1995)

$$\mathbf{v}_{i+1} = \arg \min_{\mathbf{v}\in\mathbb{R}^{2N}} \|\mathbf{z} - \mathbf{J}_i(2\mathbf{v} - \mathbf{v}_i)\|_1 + \frac{M}{2\mu_i}\|\mathbf{v} - \mathbf{v}_i\|_2^2, \tag{4.3}$$

where $i \in \mathbb{N}$ is the iteration index, $\mathbf{J}_i := [\mathbf{v}_i^{\mathrm{T}}\mathbf{H}_m]_{1\leq m\leq M}$ is an $M \times N$ matrix whose m-th row is $\mathbf{v}_i^{\mathrm{T}}\mathbf{H}_m$, and $\{\mu_i > 0\}$ is a pre-selected step-size sequence.

It is evident that the per-iteration subproblem (4.3) constitutes a convex quadratic program amenable to solution using conventional convex programming techniques. One possible iterative solver of (4.3) is based on the ADMM. Such an ADMM-based inner loop turns out to entail $2M + 2N$ auxiliary variables, thus requiring the update of $2M + 4N$ variables per iteration (Wang et al. 2019b).

In pursuit of a simplified solver, the next section will commence by reformulating (4.3) into a Lasso-type representation. Subsequently, we will delve into the unrolling of the ensuing double-loop prox-linear iterations (4.3), which form the foundational building blocks of our DNNs-based PSSE solver.

4.2 The Prox-Linear Net

In this section, we will introduce a DNNs-based scheme to approximate the solution of (4.2) by iteratively unrolling the double-loop prox-linear iterations. To facilitate this, let us define the vector variable $\mathbf{u}_i$ as $\mathbf{u}_i := \mathbf{J}_i(2\mathbf{v} - \mathbf{v}_i) - \mathbf{z}$ and subsequently substitute $\mathbf{u}_i$ into (4.3), yielding the expression

$$\mathbf{u}_i^* = \arg \min_{\mathbf{u}_i \in \mathbb{R}^M} \|\mathbf{u}_i\|_1 + \frac{M}{4\mu_i} \|\mathbf{B}_i(\mathbf{u}_i + \mathbf{z}) - \mathbf{v}_i\|_2^2, \tag{4.4}$$

where $\mathbf{B}_i \in \mathbb{R}^{2N \times M}$ satisfying $\mathbf{B}_i \mathbf{J}_i = \mathbf{I}_{2N \times 2N}$ denotes the pseudo inverse of $\mathbf{J}_i$. The pseudo inverse $\mathbf{B}_i$ exists because PSSE requires $M \geq 2N$ to guarantee system observability.

Once the inner optimum variable $\mathbf{u}_i^*$ is found, the next outer-loop iterate $\mathbf{v}_{i+1}$ can be readily obtained as

$$\mathbf{v}_{i+1} = [\mathbf{B}_i(\mathbf{u}_i^* + \mathbf{z}) + \mathbf{v}_i]/2, \tag{4.5}$$

following the definition of $\mathbf{u}_i$. Interestingly, (4.4) now reduces to a Lasso problem (Boyd et al. 2014), for which various celebrated solvers have been put forward, including e.g., the iterative shrinkage and thresholding algorithm (ISTA) (Boyd et al. 2014).

More precisely, using k to denote the iteration index of the inner-loop, the ISTA for solving (4.4) progresses through iterations at $k = 0, 1, \ldots,$ along the following recursion

$$\mathbf{u}_i^{k+1} = \mathrm{S}_\eta\left(\mathbf{u}_i^k - \frac{\eta M}{2\mu_i}\mathbf{B}_i^{\mathrm{T}}\left[\mathbf{B}_i\left(\mathbf{u}_i^k + \mathbf{z}\right) - \mathbf{v}_i\right]\right) = \mathrm{S}_\eta\left(\mathbf{W}_i^k \mathbf{u}_i^k + \mathbf{A}_i \mathbf{z} + \mathbf{b}_i^k\right), \tag{4.6}$$

where $\eta > 0$ is a fixed step size with coefficients

$$\mathbf{W}_i^k := \mathbf{I} - \frac{\eta M}{2\mu_i}\mathbf{B}_i^{\mathrm{T}}\mathbf{B}_i, \quad \forall k \in \mathbb{N}, \tag{4.7a}$$

$$\mathbf{A}_i := -\frac{\eta M}{2\mu_i}\mathbf{B}_i^{\mathrm{T}}\mathbf{B}_i, \tag{4.7b}$$

$$\mathbf{b}_i^k := \frac{\eta M}{2\mu_i}\mathbf{B}_i^{\mathrm{T}}\mathbf{v}_i, \quad \forall k \in \mathbb{N}, \tag{4.7c}$$

and $S_\eta(\cdot)$ is the so-called soft thresholding operator

$$S_\eta(x) := \begin{cases} x - \eta, x > \eta, \\ 0, -\eta \leq x \leq \eta, \\ x + \eta, x < -\eta, \end{cases} \tag{4.8}$$

understood entry-wise when applied to a vector. Regarding initialization, $\mathbf{u}_0^0 = \mathbf{0}$ without any loss of generality. Moreover, for $i \geq 1$, it is valid to initialize $\mathbf{u}_i^0$ as $\mathbf{u}_{i-1}^*$.

A reduced-complexity implementation of the prox-linear PSSE solver is outlined in Algorithm 4.1. By selecting appropriate step sizes μ_i and η, the sequence $\mathbf{v}_i$ generated by Algorithm 4.1 converges towards a stationary point of (4.2) (Wang et al. 2019b). It's worth noting that Algorithm 4.1 often demands a substantial number of iterations, denoted as K and I, to approximate the solution of (4.2). Additionally, the computation of the pseudo-inverse $\mathbf{B}_i$ is required in each outer-loop iteration. These challenges can discourages its use in real-time applications.

Algorithm 4.1 Reduced-complexity prox-linear solver

1: **Input** data $\{(z_m, \mathbf{H}_m)\}_{m=1}^{M}$, step sizes $\{\mu_i\}$, η, and initialization $\mathbf{v}_0 = \mathbf{1}$, $\mathbf{u}_0^0 = \mathbf{0}$
2: **For** $i = 0, 1, \ldots, I$ **do**
3: **Evaluate** $\mathbf{W}_i^k$, $\mathbf{A}_i$, and $\mathbf{b}_i^k$ according to (4.7a), (4.7b) and (4.7c)
4: **Initialize** $\mathbf{u}_i^0$
5: **For** $k = 0, 1, \ldots, K$ **do**
6: **Update** $\mathbf{u}_i^{k+1}$ using (4.6)
7: **End for**
8: **Update** $\mathbf{v}_{i+1}$ using (4.5)
9: **End for**

Rather than relying on solving the optimization problem through a potentially extensive number of iterations, recent approaches (e.g., Gregor and LeCun 2010; Sun et al. 2016) advocate the use of trainable DNNs constructed by unfolding and truncating these iterations. This innovative approach allows for data-driven solutions. Specifically, the matrices $\{\mathbf{W}_i^k\}$ and vectors $\{b_i^k\}$ are learned from historical data through back-propagation, as opposed to being determined based on a fixed criterion like (4.7a), (4.7b) and (4.7c). Consequently, well-trained unrolled DNNs can achieve competitive performance even with a relatively small number of 'iterations', referred to as layers in this context; for more insights, see also (Gregor and LeCun 2010; Wang et al. 2016; Sun et al. 2016). In Sect. 4.4, we will demonstrate the effectiveness of our unrolled prox-linear network by showcasing a significant speedup compared to conventional optimization-based PSSE solvers. Remarkably, this speedup is achieved while requiring only a limited number of matrix–vector multiplications. In the subsequent sections, we provide a detailed illustration of how Algorithm 4.1 can be unrolled to design DNNs tailored for real-time PSSE.

Consider first unrolling the outer loop (4.3) up to, say the $(I + 1)$-st iteration to obtain $\mathbf{v}_{I+1}$. Leveraging the recursion (4.6), each inner loop iteration i refines the initialization $\mathbf{u}_i^0 = \mathbf{u}_{i-1}^K$ to yield after K inner iterations $\mathbf{u}_i^K$. Such an unrolling leads to a $K(I + 1)$-layer structured DNNs. Suppose that the sequence $\{\mathbf{v}_i\}_{i=0}^{I+1}$ has converged, which means $\|= \mathbf{v}_I - \mathbf{v}_{I+1}\| \leq \varepsilon$ for some $\epsilon > 0$. It can then be deduced that $\mathbf{v}_{I+1} = \mathbf{B}_I^u \mathbf{u}_I^* + \mathbf{B}_I^z \mathbf{z}$ with $\mathbf{B}_I^u := \mathbf{B}_I$ and $\mathbf{B}_I^z := \mathbf{B}_I$.

Our innovative DNNs architecture, which builds upon the physics-based Algorithm 4.1, represents a hybrid approach that combines plain-vanilla FNNs with a conventional iterative solver such as Algorithm 4.1. We shall henceforth refer to it as the 'prox-linear net'. To illustrate, a prox-linear net with $K = 3$ is presented in Fig. 4.2. The first inner loop, $i = 0$, is highlighted within a dashed box, where $\mathbf{u}_0^1 = S\eta(\mathbf{A}_0\mathbf{z} + \mathbf{b}_0)$ since $\mathbf{u}_0^0 = \mathbf{0}$. As with (Gregor and LeCun 2010; Wang et al. 2016; Sun et al. 2016), our prox-linear net can be treated as a trainable regressor that predicts $\mathbf{v}$ from data $\mathbf{z}$, where the coefficients $\{\mathbf{b}_i^k\}_{0\leq i\leq I}^{1\leq k\leq 3}$, $\{\mathbf{A}_i\}_{i=0}^{I}$, $\{\mathbf{W}_i^k\}_{0\leq i\leq I}^{1\leq k\leq 3}$, $\mathbf{B}_I^u$, and

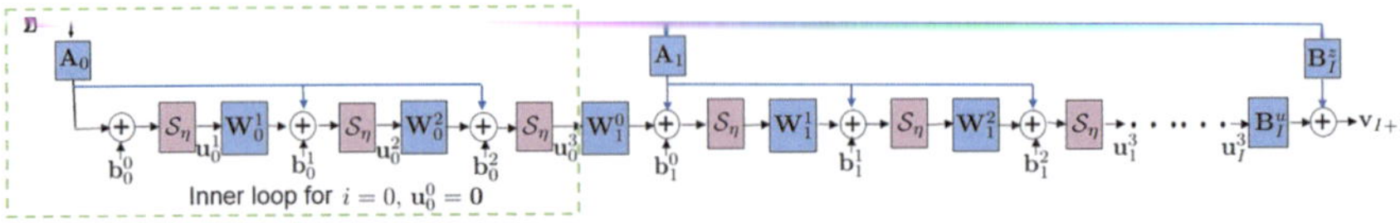

Fig. 4.2 Prox-linear net with K = 3 blocks

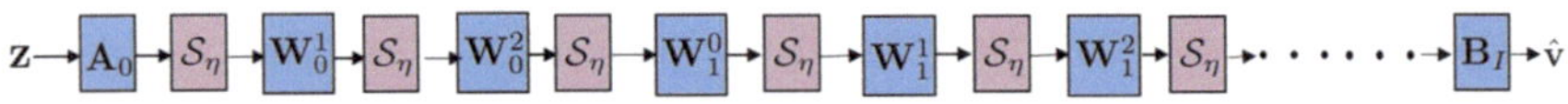

Fig. 4.3 Plain-vanilla FNNs which has the same per-layer number of hidden units as the prox-linear net

$\mathbf{B}_I^z$ are typically untied to enhance approximation capability and learning flexibility. Given a dataset of historical and/or simulated measurement-voltage training pairs $\{(\mathbf{z}_s, \mathbf{v}_s)\}$, these coefficients can be learned in an end-to-end manner using back-propagation (Rumelhart et al. 1986). Additionally, employing the (Huber loss Huber 2004) during training can impart resilience to outliers in the state estimates.

In comparison to the conventional FNNs illustrated in Fig. 4.3, our proposed prox-linear net introduces two key enhancements: (i) Skip-connections (depicted by the bluish lines in Fig. 4.2) that directly link the input **z** to intermediate and output layers. (ii) A fixed number (M in this case) of hidden neurons per layer. Both analytically and empirically, these skip-connections have been demonstrated to mitigate the issues of 'vanishing' and 'exploding' gradients, thus facilitating successful and efficient training of DNNs (He et al. 2016a, 2016b). Furthermore, 'skip-connections' play a pivotal role in endowing DNNs with universal approximation capabilities when a fixed number of hidden neurons per layer are employed (Lin and Jegelka 2018).

The only hyper-parameters requiring tuning in our prox-linear net are I, K, and η, which also function as tuning parameters in the iterative optimization solver presented in Algorithm 4.1. Notably, in addition to the soft-thresholding nonlinearity, as illustrated in Figs. 4.2 and 4.3, alternative functions such as the rectified linear unit (ReLU) can be utilized (Gregor and LeCun 2010; Wang et al. 2019a). In our simulated tests, we have observed that both the prox-linear net with soft-thresholding operators and the one with ReLUs yield comparable performance. By default, we employ ReLU activation functions to comprehensively explore how different network architectures impact performance, unless explicitly specified otherwise.

The flow chart illustrating the prox-linear net for real-time PSSE is presented in Fig. 4.4. The real-time inference stage is described within the left rounded rectangular box, while the offline training stage is depicted on the right. This seamless integration of the physics described in (4.1) with our DNNs architecture design is a pivotal feature. The extensive numerical tests conducted in Sect. 4.4 will unequivocally substantiate the remarkable performance enhancement achieved by our prox-linear nets in comparison to competing FNNs and Gauss–Newton based PSSE approaches.

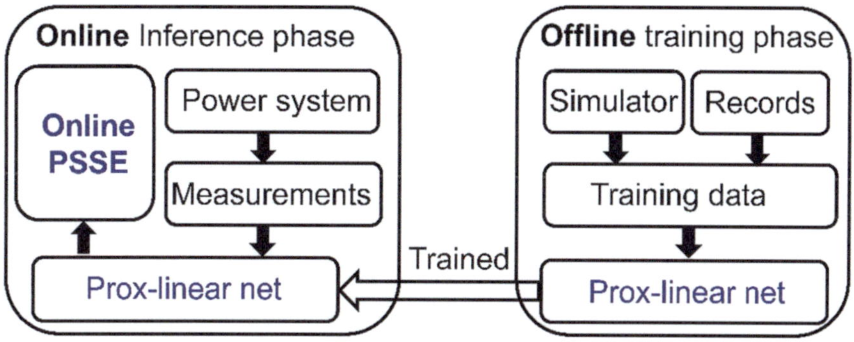

Fig. 4.4 Deep prox-linear net based real-time PSSE

4.3 Deep Recurrent Neural Networks for State Forecasting

Per time slot t, the PSSE scheme, as detailed in Sect. 4.2, is designed to estimate the state vector $\mathbf{v}_t \in \mathbb{R}^{2N}$ based on the received measurements $\mathbf{z}_t$. However, this scheme faces a notable challenge when dealing with missing entries in $\mathbf{z}_t$, a common occurrence in a SCADA system due to factors such as meter failures or communication breakdowns (Kekatos et al. 2017). To further enhance the capabilities of our innovative PSSE approach and gain insights into the power system's future state, we are motivated to explore power system state forecasting. In this context, state forecasting involves predicting the next state, $\mathbf{v}_{t+1}$, at time slot $t+1$, given the available time-series data $\{\mathbf{v}_\tau\}_{\tau=0}^{t}$ (Do Coutto Filho and Souza 2009). Analytically, the estimation and prediction steps can be outlined as follows

$$\mathbf{v}_{t+1} = \phi(\mathbf{v}_t, \mathbf{v}_{t-1}, \mathbf{v}_{t-2}, \ldots, \mathbf{v}_{t-r+1}) + \boldsymbol{\xi}_t, \tag{4.9}$$

$$\mathbf{z}_{t+1} = \mathbf{h}_{t+1}(\mathbf{v}_{t+1}) + \varepsilon_{t+1}, \tag{4.10}$$

where $\{\boldsymbol{\xi}_t, \epsilon_{t+1}\}$ account for modeling inaccuracies; the tunable parameter $r \geq 1$ represents the number of lagged (present included) states used to predict $\mathbf{v}_{t+1}$; and the unknown (nonlinear) function ϕ captures the state transition, while $\mathbf{h}_{t+1}(\cdot)$ is the measurement function that summarizes equations in (4.1) at time slot $t+1$. To accomplish state forecasting effectively, we must address the estimation or approximation of function ϕ. In the forthcoming sections, we will introduce an approach that leverages RNNs for this purpose.

RNNs are a class of NNs models specially designed for learning from correlated time series data. In comparison to FNNs, RNNs offer distinct advantages. They exhibit the ability to handle long-memory inputs, where input sequences involve extensive historical data with a large r. Additionally, RNNs have the ability to process input sequences of varying lengths, making them particularly versatile and adaptable in handling dynamic data structures (Goodfellow et al. 2016). Given the input sequence $\{\mathbf{v}_\tau\}_{\tau=t-r+1}^{T}$, and an initial state $\mathbf{s}_{t-r}$, an RNNs finds the hidden state vector sequence $\{\mathbf{s}_{t-r}\}_{\tau=t-r+1}^{t}$ by repeating

$$\mathbf{s}_\tau = f(\mathbf{R}^0\mathbf{v}_\tau + \mathbf{R}^{ss}\mathbf{s}_{\tau-1} + \mathbf{r}^0), \tag{4.11}$$

where $f(\cdot)$ represents a nonlinear activation function (e.g., ReLU or sigmoid unit), which is applied entry-wise to a vector. Meanwhile, the coefficient matrices $\mathbf{R}^0$ and $\mathbf{R}^{ss}$, along with the vector $\mathbf{r}^0$, consist of time-invariant weights.

Deep RNNs, characterized by the presence of multiple layers (typically three or more) in their architecture, offer the capacity to acquire concise representations of time series data by means of hierarchical nonlinear transformations. The application of deep RNNs models has led to substantial advancements in various domains reliant on sequence processing, such as music prediction and machine translation (Goodfellow et al. 2016). By stacking up multiple recurrent hidden layers (cf. (4.11)) one on top of another, deep RNNs can be constructed as follows (Pascanu et al. 2013).

$$\mathbf{s}_\tau^l = f(\mathbf{R}^{l-1}\mathbf{s}_\tau^{l-1} + \mathbf{R}^{ss,l}\mathbf{s}_{\tau-1}^l + \mathbf{r}^{l-1}), \quad l \geq 1, \tag{4.12}$$

where l is the layer index, $\mathbf{s}_\tau^l$ denotes the hidden state of the l-th layer at slot τ having $\mathbf{s}_\tau^0 := \mathbf{v}_\tau$, and $\{\mathbf{R}^l, \mathbf{R}^{ss,l}, \mathbf{r}^l\}$ collect all unknown weights. Figure 4.5 (left) depicts the computational graph representing (4.12) for $l = 2$, with the bias vectors $\mathbf{r}^l = \mathbf{0}$, $\forall l$ for simplicity in depiction, and the black squares standing for single-step delay units. Upon unfolding the graph by eliminating loops and linking the arrows to their respective units in the subsequent time slot, we obtain a deep RNN, as depicted in Fig. 4.5 (right). Here, the rows correspond to layers, while the columns signify individual time slots.

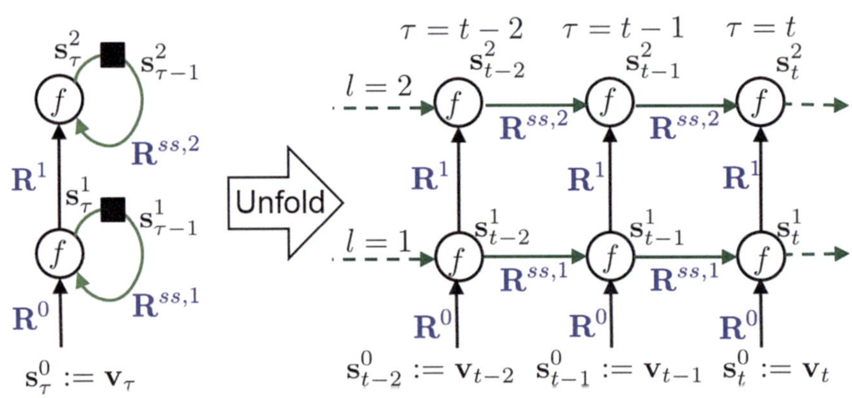

Fig. 4.5 An unfolded deep RNNs with no outputs

The output of an RNN can manifest in different ways, encompassing single outputs at per time step or a solitary output after multiple steps. The latter situation corresponds to the notion of r th-order nonlinear regression, as illustrated in (4.9), where we approximate ϕ through a deep RNN. To be more specific, the output of our deep RNN can be formulated as follows

$$\breve{\mathbf{v}}_{t+1} = \mathbf{R}^{out}\mathbf{s}_t^l + \mathbf{r}^{out}, \tag{4.13}$$

where $\breve{\mathbf{v}}_{t+1}$ is the forecast of $\mathbf{v}_{t+1}$ at time t, and $(\mathbf{R}^{out}, \mathbf{r}^{out})$ contain weights of the output layer. Given historical voltage time series, it is possible to learn the weights $(\mathbf{R}^{out}, \mathbf{r}^{out})$ along with $\{\mathbf{R}^l, \mathbf{R}^{ss,l}, \mathbf{r}^l\}$ through an end-to-end learning process utilizing backpropagation (Good-fellow et al. 2016). By employing RNNs in the context of state-space models, the class of nonlinear predictors discussed in (Do Coutto Filho and Souza 2009) is significantly expanded to incorporate memory. As will become evident from our extensive numerical experiments, the performance of our forecasting models can be notably enhanced by leveraging deep RNNs. It is worth noting that while our primary focus is on one-step state forecasting, it's important to emphasize that our proposed methodologies can be readily adapted with minor adjustments to predict system states multiple steps ahead.

Up to this point, we have discussed the utility of RNNs in creating adaptable nonlinear predictors for power system state forecasting. To predict $\breve{\mathbf{v}}_{t+1}$ at time slot t, the RNNs model presented in (4.12) relies on having access to the ground-truth voltages $\{\mathbf{v}_\tau\}_{\tau=t-r+1}^{\mathrm{T}}$ (cf. (4.9)). However, in practical cases, these ground-truth values may not always be readily available. In such cases, we can employ the estimated voltages $\{\hat{\mathbf{v}}_\tau\}_{\tau=t-r+1}^{\mathrm{T}}$ provided by our prox-linear net-based estimator discussed in Sect. 4.2. Subsequently, the forecasted state $\breve{\mathbf{v}}_{t+1}$ can serve as a prior to assist the PSSE process at time slot $t+1$. This assistance is achieved by generating virtual measurements, referred to as $\breve{\mathbf{z}}_{t+1}$, through the measurement function $\mathbf{h}_{t+1}(\breve{\mathbf{v}}_{t+1})$. These virtual measurements can be readily incorporated into (4.2) to enhance the PSSE performance. For instance, in situations involving missing entries in $\mathbf{z}_{t+1}$, the $\breve{\mathbf{z}}_{t+1}$ values can be employed to improve PSSE accuracy by imputing the missing data points.

Figure 4.6 presents the flow chart illustrating the comprehensive real-time power system monitoring scheme. This scheme encompasses both the deep prox-linear net-based PSSE module, as shown in Fig. 4.1, and the deep RNNs-based state forecasting module, as depicted in Fig. 4.4. These modules operate concurrently at time t and $t+1$ to provide a holistic understanding of the power system's status. In our experimental setup, we utilized RNNs with $l=3$ and $r=10$. Our novel approach bears resemblance to predictor–corrector-type estimation methods commonly employed in dynamic SE problems, often implemented using Kalman filters (Kekatos et al. 2017). It is worth noting that, while not explored in depth within this section, the

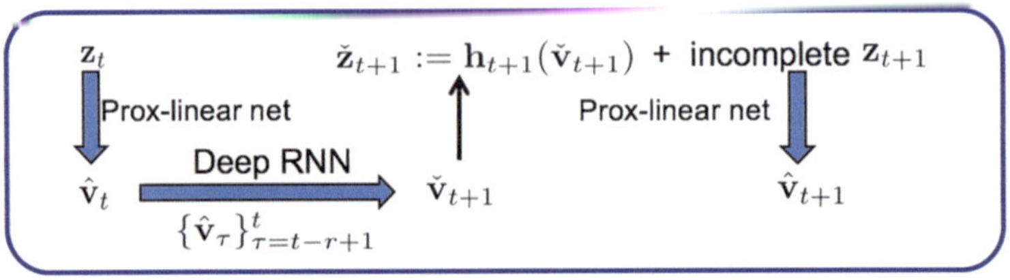

Fig. 4.6 DNNs-based real-time power system monitoring

discrepancies between the actual measurements $\mathbf{z}_{t+1}$ and both the PSSE-based estimates $\mathbf{h}_{t+1}(\hat{\mathbf{v}}_{t+1})$ and the RNNs-based predictions $\check{\mathbf{z}}_{t+1}$ can serve as valuable indicators. These discrepancies can help identify erroneous data, alterations in the grid topology, and changes in the link parameters, as elaborated in (Do Coutto Filho and Souza 2009).

4.4 Numerical Tests

The performance of our deep prox-linear net based PSSE, and deep RNNs based state forecasting methods was evaluated using the IEEE 57- and 118-bus benchmark systems. Real load data from the 2012 Global Energy Forecasting Competition was used to generate the training and testing datasets, where the load series were subsampled for size reduction by a factor of 5 (2) for the IEEE 57-bus (118-bus) system. Subsequently, the resultant load instances were normalized to match the scale of power demands in the simulated system. The MATPOWER toolbox (Zimmerman et al. 2010) was used to solve the AC power flow equations with the normalized load series as inputs, to obtain the ground-truth voltages $\{\mathbf{v}_\tau\}$, and produce measurements $\{\mathbf{z}_\tau\}$ that comprise all forwarding-end active (reactive) power flows, as well as all voltage magnitudes. All NNs were trained using 'TensorFlow' (Abadi et al. 2016) on an NVIDIA Titan X GPU with 12 GB RAM, with weights learned by the backpropagation based algorithm 'Adam' (with starting learning rate 10^{-3}) for 200 epochs. To alleviate randomness in the obtained weights introduced by the training algorithms, all NNs were trained and tested independently for 20 times, with reported results averaged over 20 runs. For reproducibility, the 'Python'-based implementation of our prox-linear net for PSSE of the 118-bus system is publicly available at https:// github.com/ LiangZhang\\UMN/PSSE-via-DNNs.

4.4.1 Prox-Linear Nets for PSSE

We commenced our evaluation by subjecting the prox-linear network-based PSSE to rigorous testing. This PSSE method aimed to estimate the state vector $\{\hat{\mathbf{v}}_\tau\}$ based on the available measurements $\{\mathbf{z}_\tau\}$. Throughout both the training and testing phases of our experiments, all measurements $\{\mathbf{z}_\tau\}$ were intentionally perturbed by additive white Gaussian noise. Specifically, the standard deviation for power flows was set to 0.02, while for voltage magnitudes, it was 0.01. To assess the performance of our prox-linear network, we employed the normalized RMSE as the evaluation metric. The normalized RMSE was calculated as $\|\hat{\mathbf{v}} - \mathbf{v}\|_2/N$, where $\mathbf{v}$ represents the ground truth state vector, and $\hat{\mathbf{v}}$ represents the estimate obtained through the prox-linear network. This metric provided valuable insights into the accuracy and fidelity of our PSSE approach.

In particular, the prox-linear net was simulated with $T = 2$ and $K = 3$. The 'workhorse' Gauss–Newton method, a 6-layer 'plain-vanilla' FNNs that has the same depth as our prox-linear net, and an 8-layer 'plain-vanilla' FNNs that has roughly the same number of parameters as the prox-linear net, were simulated as baselines. The number of hidden units per layer in all NNs was kept equal to the dimension of the input, that is, $57 \times 2 = 114$ for the 57-bus system and $118 \times 2 = 236$ for the 118-bus system.

In the first experiment using the 57-bus system, a total of 7676 measurement-voltage $(\mathbf{z}_\tau, \mathbf{v}_\tau)$ pairs were generated, out of which the first 6176 pairs were used for training, and the rest were kept for testing. The average performance over 20 trials, was evaluated in terms of the average RMSEs over the 1500 testing examples for the prox-linear net, Gauss–Newton, 6-layer FNNs, and 8-layer FNNs, are 3.49×10^{-4}, 3.2×10^{-4}, 6.35×10^{-4}, and 9.02×10^{-4}, respectively. These numbers showcase competitive performance of the prox-linear net. Interestingly, when the number of hidden layers of 'plain-vanilla' FNNs increases from 6 to 8, the performance degrades due partly to the difficulty in training the 8-layer FNNs.

As far as the computation time is concerned, the prox-linear net, Gauss–Newton, 6-layer FNNs, and 8-layer FNNs over 1500 testing examples are 0.0973s, 14.22s, 0.0944s, and 0.0954s, resulting in an average per-instance estimation time of 9.0×10^{-5} s, 4.9×10^{-2} s, 7.8×10^{-5} s, and 8.9×10^{-5} s, respectively.

These numbers corroborate the speedup advantage of NN-based PSSE over the traditional Gauss–Newton approach. The ground-truth voltages along with the estimates found by the prox-linear net, 6-layer FNNs, and 8-layer FNNs for bus 10 and bus 27 from test instances 100 to 120, are shown in Figs. 4.7 and 4.8, respectively. The true voltages and the estimated ones by NNs for all buses on test instance 120 are depicted in Fig. 4.9. Evidently, our prox-linear net based PSSE performs the best in all cases.

The second experiment tested our prox-linear net using the IEEE 118-bus system, where 18,528 voltage-measurement pairs were simulated, with 14,822 pairs employed for training and 3706 kept for testing. The average RMSEs over 3706 testing examples for the prox-linear net, Gauss–Newton, 6-layer FNNs and 8-layer FNNs, are 2.97×10^{-4}, 4.71×10^{-2}, 1.645×10^{-3}, and 2.366×10^{-3}, respectively. Clearly, the prox-linear net stands out with a substantially improved performance compared to its competitors, particularly as the system's size increases. The Gauss–Newton approach, in particular, exhibits the weakest performance due to the unbalanced grid parameters inherent in this specific test system. It is worth noting

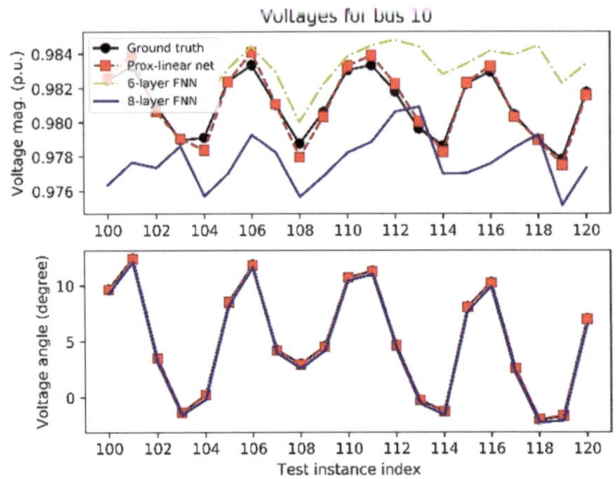

Fig. 4.7 Estimation errors in voltage magnitudes and angles of bus 10 of the 57-bus system from test instances 100 to 120

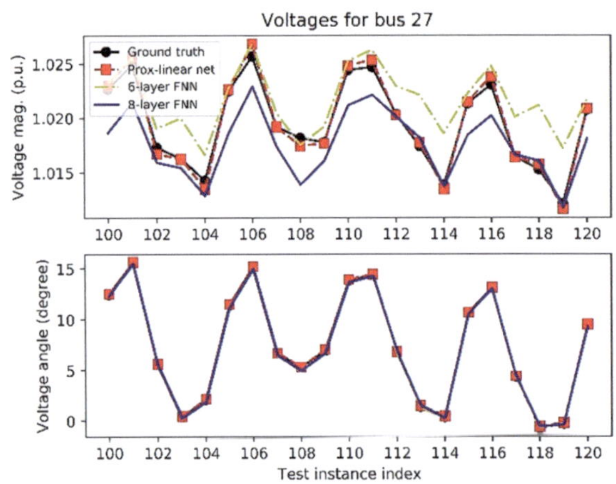

Fig. 4.8 Estimation errors in voltage magnitudes and angles of bus 27 of the 57-bus system from test instances 100 to 120

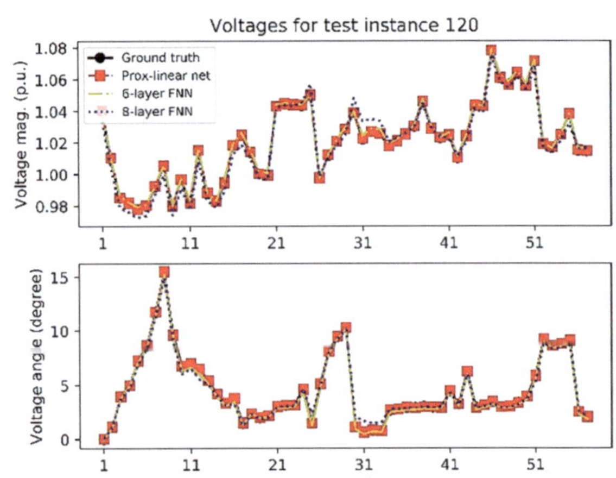

Fig. 4.9 Estimation errors in voltage magnitudes and angles of all the 57 buses of the 57-bus system at test instance 120

an interesting phenomenon observed during this experiment. In several instances, the Gauss–Newton iterations succeeded in minimizing the WLS objective function, resulting in a loss smaller than 10^{-6}. However, these iterations often converged to a stationary point that deviated significantly from the simulated ground-truth voltage values. This behavior can be attributed to the non-convex nature of the WLS function, which leads to the existence of multiple optimal solutions. Depending on the initial conditions, traditional optimization-based solvers like Gauss–Newton may unfortunately become trapped at one of these points. In stark contrast, data-driven NNs-based approaches elegantly circumvent this challenge, showcasing their robustness and superiority in handling complex and non-convex optimization landscapes.

Regarding computational efficiency, it is evident that the prox-linear net, Gauss–Newton, 6-layer FNNs, and 8-layer FNNs, when tested on 3706 examples, exhibit runtimes of 0.3323 s, 183.4 s, 0.2895 s, and 0.3315 s, respectively. These results translate to an average per-instance estimation time of 6.5×10^{-5} seconds, 9.48×10^{-3} seconds, 6.3×10^{-5} seconds, and 6.4×10^{-5} seconds, respectively. Once again, this substantiates the computational efficiency of NNs-based approaches in comparison to traditional methods. For a visual representation of the estimation quality, we present the ground-truth voltage values alongside their respective estimates derived from the prox-linear net, the 6-layer FNNs, and the 8-layer FNNs. This comparison is conducted for both bus 50 and bus 100 across test instances 1000 to 1050, as depicted in Figs. 4.10 and 4.11, respectively. Furthermore, we offer visual representations of actual voltage values and their respective estimates for the initial fifty buses in test instance 1000, as illustrated in Fig. 4.12. In each case, it becomes apparent that our prox-linear network consistently excels in terms of estimation quality compared to alternative methods.

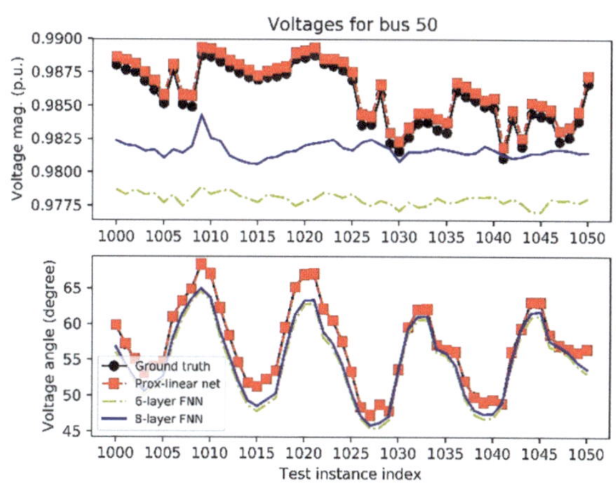

Fig. 4.10 Estimation errors in voltage magnitudes and angles of bus 50 of the 118-bus system from instances 1000 to 1050

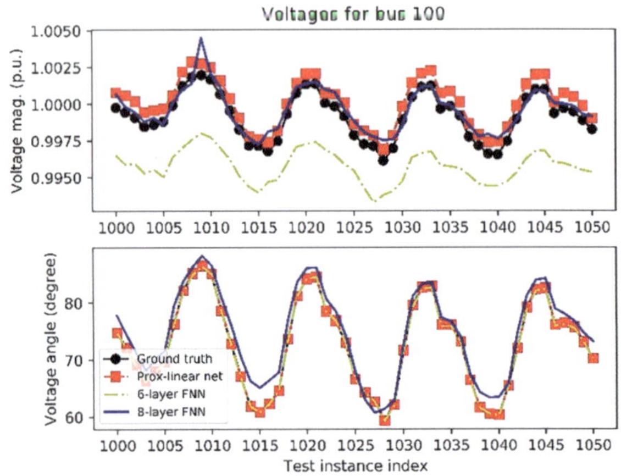

Fig. 4.11 Estimation errors in voltage magnitudes and angles of bus 100 of the 118-bus system from instances 1000 to 1050

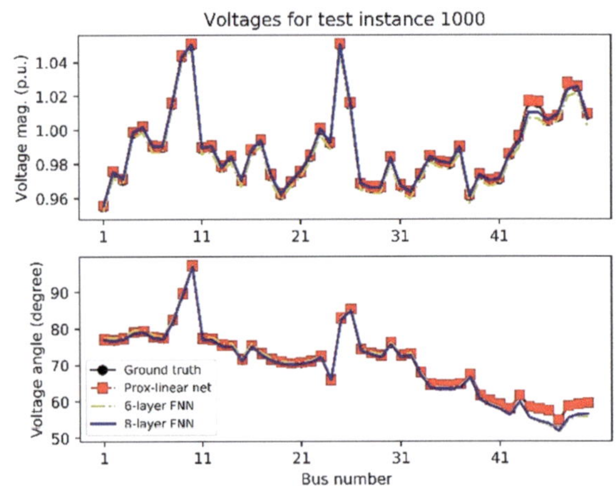

Fig. 4.12 Estimation errors in voltage magnitudes and angles of the first 50 buses of the 118-bus system at test instance 1000

4.4.2 Deep RNNs for State Forecating

This section delves into our power system state forecasting scheme based on RNNs. We assessed the forecasting performance using the normalized RMSE, calculated as $\left\|\breve{\mathbf{v}} - \mathbf{v}\right\|_2/N$, where $\breve{\mathbf{v}}$ represents the forecast and $\mathbf{v}$ is the ground truth.

In our experimental setup, deep RNNs with hyperparameters $l = 3$, $r = 10$, a batch size of 32, and ReLU activation functions were trained and tested. These RNNs were evaluated on two types of input data: the ground-truth voltage time series and the estimated voltage time series obtained from the prox-linear net. Hereafter, we will refer to the latter as RNNs with estimated voltages. The number of hidden units per layer in the RNNs was set equal to the input dimension, specifically $57 \times 2 = 114$ for the 57-bus system and $118 \times 2 = 236$ for the 118-bus system. For comparison purposes, we included two benchmark methods: a single-hidden-layer FNNs with

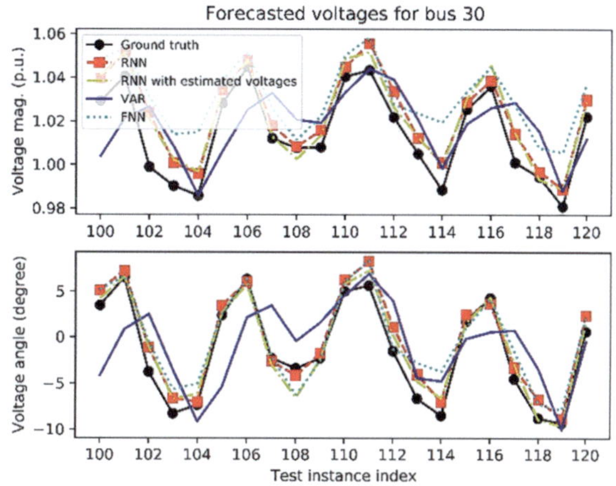

Fig. 4.13 Forecasting errors in voltage magnitudes and angles of bus 30 of the 57-bus system from test instances 100 to 120

two layers in total (2-layer FNNs) (Do Coutto Filho et al. 2009), and a vector auto-regressive (VAR) model of order 1 (Hassanzadeh et al. 2015), which were trained and tested using ground-truth voltage time series. These benchmarks allow us to assess the relative performance of our deep RNNs. The average RMSEs over 20 Monte Carlo runs for the RNNs, RNNs with estimated voltages, 2-layer FNNs, and VAR (1) are respectively 2.303×10^{-3}, 2.305×10^{-3}, 3.153×10^{-3}, and 6.772×10^{-3} for the 57-bus system, as well as 2.588×10^{-3}, 2.751×10^{-3}, 4.249×10^{-3}, 6.461×10^{-3} for the 118-bus system. These results highlight that our deep RNNs model trained with estimated voltages offers forecasting performance comparable to that achieved using ground-truth voltages. Interestingly, both the 2-layer FNNs and VAR (1) which were trained and tested with ground-truth voltage time series, perform even worse than our RNNs trained with estimated voltages.

The figures below present the actual voltage values and their corresponding forecasts by different methods for the 57-bus and 118-bus systems: For the 57-bus system, Figs. 4.13 and 4.14 display the true voltage values and forecasts for bus 30 from test instances 100–120, as well as all buses on test instance 100, respectively. For the 118-bus system, Fig. 4.15 exhibits the ground-truth and forecasted voltage values for the first 50 buses on test instance 1000. In all cases, it is evident from the curves that our deep RNNs-based approaches consistently outperform the other methods in forecasting voltage values.

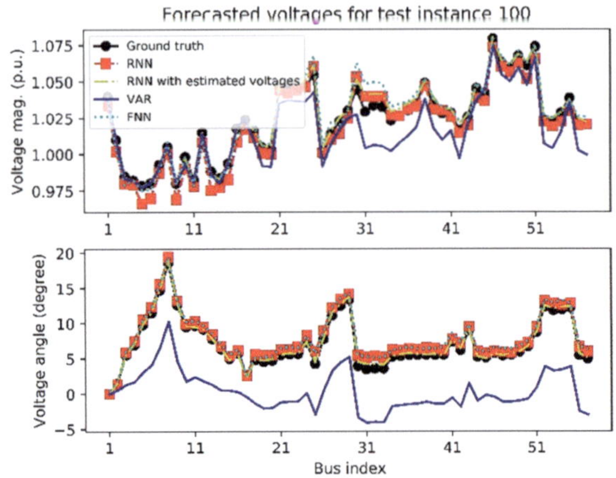

Fig. 4.14 Forecasting errors in voltage magnitudes and angles of all the 57 buses of the 57-bus system at test instance 100

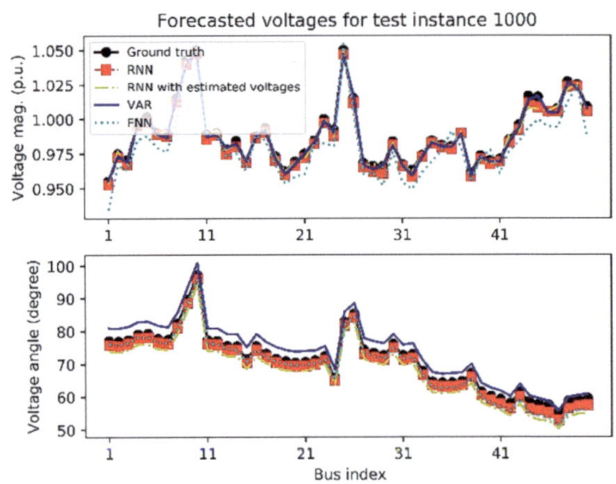

Fig. 4.15 Forecasting errors in voltage magnitudes and angles of the first 50 buses of the 118-bus system at instance 1000

Chapter 5
Data-Driven Graph Priors for State Estimation

Since state variables pose challenges for direct measurement, the SCADA system provides a wealth of measurements, encompassing voltage magnitudes, power flows, and power injections. With these SCADA measurements at hand, the primary objective of PSSE is to deduce the state variables, specifically the complex voltages at all network buses. To circumvent the challenges posed by nonconvex optimization in power system monitoring and control, recent studies have shifted their focus toward the development of data-driven and model-driven NNs solutions, as documented in references (Barbeiro et al. 2014; Manitsas et al. 2012; Zhang et al. 2019b; Zamzam and Sidiropoulos 2020; Ostrometzky et al. 2019; Yang et al. 2020a; Tran et al. 2021; Yang et al. 2020b; Singh et al. 2021; Yang et al. 2019).

Extending the insights from references (Lingala and Jacob 2013; Aggarwal et al. 2018) to the realm of power systems, this chapter proposes the adoption of a trainable DNNs prior for conventional ill-posed PSSE. This approach is designed to facilitate the derivation of physically meaningful solutions in the context of PSSE. In practical applications, the measurements obtained through the SCADA system can be significantly compromised, primarily attributable to factors such as parameter uncertainty, instrument mis-calibration, and unobserved topology modifications, as discussed in references (Wang et al. 2019b; Constante-Flores et al. 2022). Furthermore, recent studies have highlighted the susceptibility of NNs to significant degradation in performance due to adversarial attacks, as demonstrated in reference (Kurakin et al. 2018).

To overcome these problems, this chapter draws on recent advances in the field of image denoising to propose a new PSSE framework. The framework introduces a data-driven regularisation term to capture a prior information from DNNs. Based on this regularised PSSE objective, a Gauss–Newton-type alternating minimization solver is first developed. To accommodate real-time monitoring, a new end-to-end DNNs is subsequently constructed by unfolding the proposed alternating minimization solver. The deep PSSE architecture further utilises a prior based graph neural networks (GNNs) for better understanding of the grid topology. An adversarial DNNs

G. Wang et al., *Data-driven Optimization and Control for Autonomous Energy Systems*,
https://doi.org/10.1007/978-981-95-1782-4_5

training method is also proposed to enhance the robustness of the physics based DNNs against bad data. Numerical tests using real load data on the IEEE 118-bus benchmark system demonstrate that the scheme achieves significant improvements in estimation performance and robustness in comparison to several state-of-the-art alternatives.

5.1 System Modeling and Problem Formulation

Consider an electrical grid comprising N buses (nodes) interconnected by E lines (edges), which can be conveniently represented as a graph denoted by $G := (N, E, \mathbf{W})$. Here, $N := \{1, \ldots, N\}$ encompasses all the buses, $E := \{(n, n')\} \subseteq N \times N$ accounts for the set of lines connecting these buses, and $\mathbf{W} \in \mathbb{R}^{N\times N}$ represents a weight matrix. In this matrix, the (n, n')-th element, denoted as $[\mathbf{W}]_{nn'}$, serves as a modeling parameter reflecting the impedance associated with the connection between buses n and n'. In particular, if $(n, n') \in E$, then $[\mathbf{W}]_{nn'} = w_{nn'}$; and $[\mathbf{W}]_{nn'} = 0$ otherwise. For each bus $n \in N$, let $V_n := v_n^r + jv_n^i$ be its complex voltage with magnitude denoted by $|V_n|$, and $P_n + jQ_n$ its complex power injection. For reference, collect the voltage magnitudes, active and reactive power injections across all buses into the N-dimensional column vectors $|\mathbf{V}|$, $\mathbf{P}$, and $\mathbf{Q}$, respectively.

The system state variables, denoted as $\mathbf{v} := [v_1^r\, v_1^i \ldots v_N^r\, v_N^i]^{\mathrm{T}} \in \mathbb{R}^{2N}$, encompass a comprehensive set of SCADA measurements. These measurements include voltage magnitudes, active and reactive power injections, as well as active and reactive power flows. To denote the specific locations where meters of each type are installed, we introduce the sets S_V, S_P, S_Q, E_P, and E_Q. For a more concise representation, we consolidate all measurements from these meters into a single vector $\mathbf{z} := [\{|V_n|^2\}_{n\in S_V}, \{P_n\}_{n\in S_P}, \{Q_n\}_{n\in S_Q}, \{P_{nn'}\}_{(n,n')\in E_P}, \{Q_{nn'}\}_{(n,n')\in E_Q}]^{\mathrm{T}} \in \mathbb{R}^M$. Additionally, we can express the m-th entry of $\mathbf{z}$, denoted as $\{z_m\}_{m=1}^M$, using the following model

$$z_m = h_m(\mathbf{v}) + \varepsilon_m, \forall m = 1, \ldots, M, \tag{5.1}$$

where $h_m(\mathbf{v}) = \mathbf{v}^{\mathrm{T}}\mathbf{H}_m\mathbf{v}$ for some symmetric measurement matrix $\mathbf{H}_m \in \mathbb{R}^{2N\times 2N}$, and ϵ_m captures the modeling error as well as the measurement noise.

The objective of PSSE is to accurately reconstruct the state vector $\mathbf{v}$ based on the available measurements $\mathbf{z}$. To achieve this, we employ a least-squares criterion and vectorize the terms in (5.1). This formulation transforms PSSE into a Nonlinear Least-Squares (NLS) problem, given by the following equation

$$\mathbf{v}^* := \arg \min_{\mathbf{v}\in\mathbb{R}^{2N}} \|\mathbf{z} - h(\mathbf{v})\|^2. \tag{5.2}$$

Several algorithms have been developed to tackle the optimization problem defined in (5.2). These methods encompass well-established techniques like the Gauss–Newton iterations, as well as more recent approaches based on convex relaxation utilizing SDP (Zhu and Giannakis 2014). These algorithms typically commence with an initial estimate $\mathbf{v}_0$ and then iteratively apply a mapping from $\mathbf{v}_i$ to $\mathbf{v}_{i+1}$. This iterative process aims to produce a sequence of iterates that, ideally, converges to the true solution $\mathbf{v}^*$ or a point in its vicinity. In the next subsection, we will discuss the Gauss–Newton PSSE solver, a commonly used and important method in this context.

5.1.1 Gauss–Newton Iterations

The Gauss–Newton method is the most commonly used one for minimizing NLS (Bertsekas 1997). It relies on Taylor's expansion to linearize the function $h(\mathbf{v})$. Specifically, at a given point $\mathbf{v}_i$, it linearly approximates.

$$\tilde{\mathbf{H}}(\mathbf{v}, \mathbf{v}_i) \approx \mathbf{H}(\mathbf{v}_i) + \mathbf{J}_i\ (\mathbf{v} - \mathbf{v}_i), \tag{5.3}$$

where $\mathbf{J}_i := \nabla \mathbf{H}(\mathbf{v}_i)$ is the $M \times 2N$ Jacobian of h evaluated at $\mathbf{v}_i$, with $[\mathbf{J}_i]_{m,n} := \partial \mathbf{H}_m / \partial \mathbf{v}_n$. Subsequently, the Gauss–Newton method approximates the nonlinear term $\mathbf{H}(\mathbf{v})$ in (5.2) via (5.3), and finds the next iterateas its minimizer, that is

$$\mathbf{v}_{i+1} = \arg\min_{\mathbf{v}} \|\mathbf{z} - h(\mathbf{v}_i) - \mathbf{J}_i\ (\mathbf{v} - \mathbf{v}_i)\|^2. \tag{5.4}$$

Indeed, it is evident that the subproblem (5.4) at per-iteration represents a convex quadratic optimization problem. When the matrix $\mathbf{J}_i^{\mathrm{T}} \mathbf{J}_i$ is invertible, the update for the iterate $\mathbf{v}_i$ can be computed in a closed-form expression as follows

$$\mathbf{v}_{i+1} = \mathbf{v}_i + (\mathbf{J}_i^{\mathrm{T}} \mathbf{J}_i)^{-1} \mathbf{J}_i^{\mathrm{T}} (\mathbf{z} - \mathbf{H}(\mathbf{v}_i)), \tag{5.5}$$

until some stopping criterion is satisfied. In practice however, due to the matrix inversion, the Gauss–Newton method becomes computationally expensive; it is also sensitive to initialization, and in certain cases it can even diverge. These limitations discourage its application in real-time monitoring, particularly for large-scale networks. To overcome these challenges, we explore an alternative approach that avoids the need for iterative solutions to each individual PSSE instance, which arises when dealing with new sets of measurements in $\mathbf{z}$. Instead, we propose an end-to-end approach based on DNNs.

5.2 Unrolled Gauss–Newton with Deep Priors

As mentioned earlier, PSSE can be underdetermined and thus ill posed due to e.g., lack of observability that is an emerging issue in distribution networks (Bhela et al. 2017). While the primary focus of this section is on transmission systems, with the aim of illustrating the utility and advantages of our data-driven priors in PSSE, it is essential to emphasize that the proposed approach remains readily applicable to distribution systems. In response to this challenge, this section introduces a versatile topology-aware prior, which can be seamlessly integrated into the PSSE cost function regularization as outlined in Eq. (5.2). To address the resulting regularized PSSE problem, we develop an alternating minimization-based solver. Such a novel DNNs is built using several layers of unrolled Gauss–Newton iterations followed by proximal steps to account for the regularization term. An intriguing aspect is that by incorporating a GNNs-based prior, we can harness power network topology information to enhance the PSSE process.

5.2.1 *Regularized PSSE with Deep Priors*

In practical applications, the task of recovering $\mathbf{v}$ from $\mathbf{z}$ can often become ill-posed, particularly when $\mathbf{J}_i$ takes on a rectangular matrix form. Drawing inspiration from the data-driven deep priors commonly employed in image denoising (Rudin et al. 1992; Schlemper et al. 2017; Aggarwal et al. 2018), we propose the incorporation of a trainable prior information as a regularization technique for any PSSE loss. This regularization step is applied to the NLS problem presented in (5.2)

$$\min_{\mathbf{v}\in\mathbb{R}^{2N}} \|\mathbf{z}-\mathbf{H}(\mathbf{v})\|^2 + \lambda\|\mathbf{v}-\mathcal{D}(\mathbf{v})\|^2, \tag{5.6}$$

where $\lambda \geq 0$ serves as a tuning hyper-parameter, and the regularization term encourages the system states $\mathbf{v}$ to closely align with $\mathcal{D}(\mathbf{v})$. The latter $\mathcal{D}(\mathbf{v})$ could represent a nonlinear estimator $\hat{\mathbf{v}}$, possibly obtained offline based on training data. To encompass a wide range of prior information, we propose the use of a DNNs-based estimator $\mathcal{D}_\theta(\mathbf{v})$ with trainable weights θ. These weights can be learned directly from historical training data, allowing the DNNs $\mathcal{D}_\theta(\cdot)$, from a Bayesian perspective, to ideally produce the posterior mean for a given input.

While this regularizer can effectively handle ill conditioning, it's important to note that the PSSE objective in (5.6) remains nonconvex. Furthermore, the nested structure of $\mathcal{D}_\theta(\cdot)$ introduces additional complexities. Similar to the Gauss–Newton method employed for solving the NLS problem in (5.2), we address these challenges by employing an alternating minimization algorithm This iterative approach starts with an initial guess $\mathbf{v}_0$, and at per iteration i, it utilizes a linearized data consistency term to compute the next iterate $\mathbf{v}_{i+1}$, that is

$$\mathbf{v}_{i+1} = \arg\min_{\mathbf{v}} \|\mathbf{z} - h(\mathbf{v}_i) - \mathbf{J}_i\ (\mathbf{v} - \mathbf{v}_i)\|^2 + \lambda\|\mathbf{v} - \mathcal{D}_\theta(\mathbf{v}_i)\|^2 = \mathbf{A}_i\mathbf{z} + \mathbf{B}_i\mathbf{u}_i + \mathbf{b}_i, \tag{5.7}$$

where we define

$$\mathbf{A}_i := (\mathbf{J}_i^{\mathrm{T}}\mathbf{J}_i + \lambda I)^{-1}\mathbf{J}_i^{\mathrm{T}}, \tag{5.8a}$$

$$\mathbf{B}_i := \lambda(\mathbf{J}_i^{\mathrm{T}}\mathbf{J}_i + \lambda I)^{-1}, \tag{5.8b}$$

$$\mathbf{b}_i := (\mathbf{J}_i^{\mathrm{T}}\mathbf{J}_i + \lambda I)^{-1}\mathbf{J}_i^{\mathrm{T}}(\mathbf{J}_i\mathbf{v}_i - h(\mathbf{v}_i)). \tag{5.8c}$$

The solution of (5.6) can thus be approached by alternating between the ensuing two steps

$$\mathbf{u}_i \ = \mathcal{D}_\theta(\mathbf{v}_i), \tag{5.9a}$$

$$\mathbf{v}_{i+1} \ = \mathbf{A}_i\mathbf{z} + \mathbf{B}_i\mathbf{u}_i + \mathbf{b}_i. \tag{5.9b}$$

Certainly, with an initial guess $\mathbf{v}_0 = 0$ and input $\mathbf{z}$, the first iteration results in $\mathbf{v}_1 = \mathbf{A}_0\mathbf{z} + \mathbf{B}_0\mathbf{u}_0 + \mathbf{b}_0$. Subsequently, when we pass $\mathbf{v}_1$ through the DNNs $\mathcal{D}_\theta(\cdot)$ we obtain $\mathbf{u}_1$, which serves as the input for the second iteration. This second iteration can be described as $\mathbf{u}_1 = \mathcal{D}_\theta(\mathbf{v}_1)$ [as shown in (5.9a)]. In principle, one could continue these alternating iterations whenever new measurements $\mathbf{z}$ are available. However, it's essential to note that at per iteration i, the Jacobian matrix $\mathbf{J}_i$ must be evaluated, followed by matrix inversions to compute $\mathbf{A}_i$, $\mathbf{B}_i$, and $\mathbf{b}_i$. This computational burden can be particularly prohibitive for real-time monitoring tasks involving large-scale power systems.

To facilitate rapid implementation, we pursue an end-to-end learning approach, training a DNNs that is constructed by unrolling iterations of the alternating minimizer. This DNNs is designed to directly approximate the mapping from measurements $\mathbf{z}$ to states $\mathbf{v}$. For a visual representation of the resulting architecture, please refer to Fig. 5.1. It's worth noting that, in order to derive the alternating minimizer, we initially assumed that the DNNs prior $\mathcal{D}_\theta(\cdot)$ in (5.9a) was pre-trained, with its weights θ fixed in advance. However, in our GNU-NNs approach, we consider all coefficients $\{\mathbf{A}_i\}_{i=0}^{I}$, $\{\mathbf{B}_i\}_{i=0}^{I}$, $\{\mathbf{b}_i\}_{i=0}^{I}$, as well as the DNNs weights $\{\theta_i\}_{i=0}^{I}$, to be trainable and learned from the available data.

The end-to-end GNU-NNs can be effectively trained using backpropagation, leveraging historical or simulated measurements $\{\mathbf{z}^t\}_{t=1}^{\mathrm{T}}$ along with corresponding ground-truth states $\{\mathbf{v}^{*t}\}_{t=1}^{\mathrm{T}}$. Remarkably, our GNU-NNs' training process involves just a few matrix–vector multiplications, yet it attains competitive performance in PSSE when compared to other iterative solvers like the Gauss–Newton method. Moreover, in contrast to existing data-driven NN approaches, our GNU-NNs mitigates issues related to vanishing and exploding gradients. This favorable outcome is

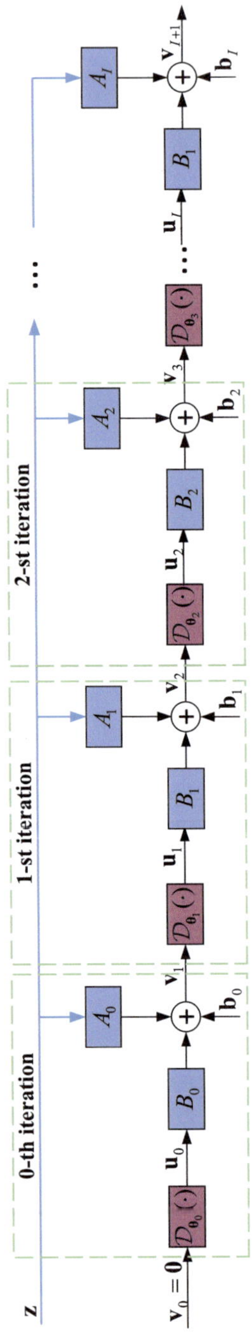

Fig. 5.1 The structure of the proposed GNU-NNs

achieved through the incorporation of direct connections, often referred to as skipping connections, which link the input layer to intermediate and output layers. These connections contribute to the network's stability and facilitate more effective training.

Remark 5.1 While it is important to acknowledge that the problem retains its non-convex nature, and iterative algorithms may occasionally converge to local solutions, the primary strength of our data-driven-based PSSE lies in its ability to harness the wealth of historical data available for training a meaningful prior. More specifically, widely employed algorithms like SGD and its variants (as discussed in, for instance, (Kingma and Ba 2014; Duchi et al. 2011; Wang et al. 2019a), have been empirically shown to frequently converge to highly satisfactory, if not near-optimal, solutions during the process of updating the NN weights. This characteristic lends robustness to our approach and underscores its practical utility. In order to mitigate challenges, such as the issue of 'overfitting' and enhance overall generalization performance, it is common practice to employ large training datasets.

Remark 5.2 In comparison to Gauss–Newton iterations, we've achieved a remarkable complexity reduction per iteration during inference, precisely by an order of magnitude, transitioning from $O(N^3)$ to $O(N^2)$. It's worth noting that for a square and invertible matrix, the computational complexity associated with Jacobian inversion is $O(N^3)$, whereas matrix–vector multiplication holds a $O(N^2)$ complexity. In our context, this substantial improvement in efficiency is made feasible by redistributing the computational load from 'online-inference' to 'offline-training' stages, utilizing historical data. Consequently, challenges such as sensitivity, hyper-parameter tuning, and convergence are predominantly addressed during the training phase. Once the mapping function between the measurement vector $\mathbf{z}$ and the state vector $\mathbf{v}$ is acquired, estimating states for a new set of measurements becomes much simpler. We just need to perform basic matrix–vector multiplications, which means passing the measurements through the trained NNs. This significant improvement in PSSE efficiency makes it much easier to achieve real-time SE.

Notably, the choice of the specific model for $\mathcal{D}_\theta(\cdot)$ offers the intriguing possibility of inducing favorable properties such as scalability and high estimation accuracy. For instance, opting for feedforward NNs as $\mathcal{D}_\theta(\cdot)$ enables the attainment of a scalable solution suitable for large power networks. However, it's essential to recognize that feedforward NNs can only indirectly incorporate grid topology information through simulated MATPOWER data. This consideration prompts us to delve into the realm of GNNs, which possess the unique capability to explicitly capture the topology and physical characteristics of the power network. The resultant Gauss–Newton unrolled with GNNs priors (GNU-GNNs) is elaborated next.

5.2.2 Graph Neural Networks Deep Prior

To facilitate the incorporation of highly expressive state estimators into our regularization term, we adopt a modeling approach for $\mathcal{D}_\theta(\cdot)$ based on GNNs, which represent a judicious choice for handling networked data. GNNs have recently showcased remarkable performance across various domains, encompassing tasks such as classification, recommendation, prediction, and robotics (Kipf and Welling 2016; Huang et al. 2021). Operating directly on graph-structured data, GNNs possess the unique capability to explicitly harness the topology of the power network. Consequently, they stand as attractive candidates for parameterization in domains where data conform to a graph structure (Kipf and Welling. 2016).

Consider a graph composed of N nodes, characterized by a weighted adjacency matrix $\mathbf{W}$ capturing node connectivity. We introduce a data matrix $\mathbf{X} \in \mathbb{R}^{N\times F}$, where each n-th row $\mathbf{X}_n^{\mathrm{T}} := [\mathbf{X}]_{n:}$ represents an $F \times 1$ feature vector associated with node n.. In the context of the PSSE problem, these features correspond to the real and imaginary parts of the nodal voltage ($F = 2$). To facilitate information propagation across the network, we perform a pre-multiplication of the input matrix $\mathbf{X}$ by $\mathbf{W}$, resulting in the diffusion of features throughout the graph. This process yields a diffused version denoted as $\breve{\mathbf{Y}} \in \mathbb{R}^{N\times F}$, which can be expressed as follows

$$\breve{\mathbf{Y}} = \mathbf{W}\mathbf{X}. \tag{5.10}$$

Remark 5.3 When modeling the propagation of features, a commonly adopted approach is to leverage the adjacency matrix or any other matrix that preserves the structural characteristics of the power network (i.e., $\mathbf{W}_{nn'} = 0$ if $(n, n') \notin E$). Examples of such matrices encompass the graph Laplacian, the random walk Laplacian, as well as their normalized variants.

The shift operation outlined in (5.10) involves a linear combination of the f-th features from all neighboring nodes to derive its propagated feature. More precisely, for bus n, the shifted feature $\breve{\mathbf{Y}}_{nf}$ can be expressed as

$$[\breve{\mathbf{Y}}]_{nf} = \sum_{i=1}^{N} [\mathbf{W}]_{ni}[\mathbf{X}]_{if} = \sum_{i\in N_n} w_{ni} x_i^f, \tag{5.11}$$

where $N_n = \{i \in N : (i, n) \in E\}$ denotes the set of neighboring buses for bus n. Undoubtedly, this interpretation yields a diffused copy or shifted version of $\mathbf{X}$ across the graph.

The 'graph convolution' operation within GNNs harnesses topological information to perform linear combinations of features, specifically

$$[\mathbf{Y}]_{nd} := [\mathcal{H} * \mathbf{X}; \mathbf{W}]_{nd} := \sum_{k=0}^{K-1} [\mathbf{W}^k \mathbf{X}]_{n:}[\mathbf{H}_k]_{:d}, \tag{5.12}$$

where $H := [\mathbf{H}_0 \cdots \mathbf{H}_{K-1}]$ with $\mathbf{H}_k \in \mathbb{R}^{F\times D}$ concatenating all filter coefficients. The intermediate matrix $\mathbf{Y} \in \mathbb{R}^{N\times D}$ contains D features per bus, while the operation $\mathbf{W}^k\mathbf{X}$

linearly amalgamates the features of buses within the k-hop neighborhood through the recursive application of the shift operator $\mathbf{W}$.

To construct a GNNs with L hidden layers, we denote the output of the $(l-1)$-st layer as $\mathbf{X}_{l-1}$, which serves as the input for the l-th layer, with l ranging from 1 to L. Initially, we set $\mathbf{X}_0 = \mathbf{X}$ as the input matrix. At each layer l, we compute the hidden $\mathbf{Y}_l \in \mathbb{R}^{N \times D_l}$ with D_l features by applying the graph convolution operation (5.12), as follows

$$[\mathbf{Y}_l]_{nd} = \sum_{k=0}^{K_l-1} [\mathbf{W}^k \mathbf{X}_{l-1}]_{n:} [\mathbf{H}_{lk}]_{:g}, \tag{5.13}$$

where $\mathbf{H}_{lk} \in \mathbb{R}^{F_{l-1} \times F_l}$ are the graph convolution coefficients for $k = 0, \ldots, K_l - 1$. At each layer l, we compute the output $\mathbf{X}_l$ by performing a graph convolution followed by a point-wise nonlinear operation $\sigma_l(\cdot)$. A common choice for $\sigma_l(t)$ is the ReLU defined as $\sigma_l(t) := \max\{0, t\}$ for $t \in \mathbb{R}$. For a visual representation of this process, please refer to Fig. 5.2. Expressing the graph convolution operation (5.13) in a more concise form, we obtain

$$\mathbf{X}_l = \sigma_l(\mathbf{Y}_l) = \sigma_l\left(\sum_{k=0}^{K_l-1} \mathbf{W}^k \mathbf{X}_{l-1} \mathbf{H}_{lk}\right). \tag{5.14}$$

The GNNs-based PSSE introduces a nonlinear functional operator denoted as $\mathbf{X}_L = \Phi(\mathbf{X}_0; \Theta, \mathbf{W})$. This operator maps the GNNs input $\mathbf{X}_0$ to voltage estimates while considering the underlying graph structure encoded by $\mathbf{W}$, through

$$\Phi(\mathbf{X}_0; \Theta, \mathbf{W}) = \sigma_L\left(\sum_{k=0}^{K_L-1} \mathbf{W}^k \left(\cdots\left(\sigma_1\left(\sum_{k=0}^{K_1-1} \mathbf{W}^k \mathbf{X}_0 \mathbf{H}_{1k}\right)\cdots\right)\right)\mathbf{H}_{Lk}\right), \tag{5.15}$$

where the parameter set Θ contains all the filter weights; that is, $\Theta := \{\mathbf{H}_{lk}, \forall l, k\}$, and also recall that $\mathbf{X}_0 = \mathbf{X}$.

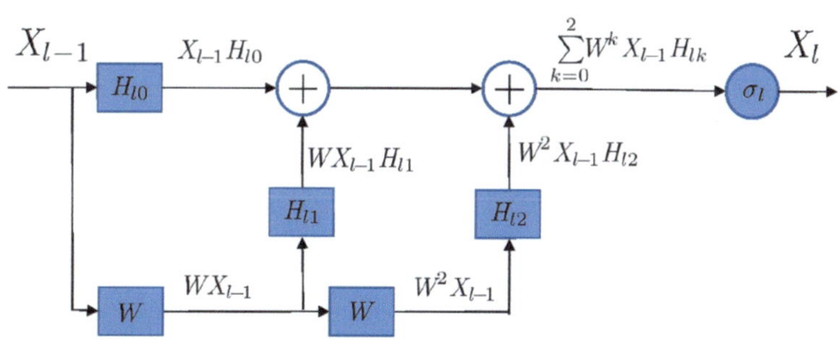

Fig. 5.2 The signal diffuses from layer $l-1$ to l with $K = 3$

Remark 5.4 With L hidden layers, each layer having F_l features and K_l filters, the total number of parameters to be learned can be calculated as $|\Theta| = \sum_{l=1}^{L} K_l \times F_l \times F_{l-1}$.

To facilitate the GNNs implementation within the proposed unrolled architecture, at the i-th iteration, we reshape the states $\mathbf{v}_i \in \mathbb{R}^{2N}$ to form the $N \times 2$ GNNs input matrix $\mathbf{X}_0^i \in \mathbb{R}^{N\times 2}$. Subsequently, we vectorize the GNNs output $\mathbf{X}_L^i \in \mathbb{R}^{N\times 2}$ to obtain the vector $\mathbf{u}_i \in \mathbb{R}^{2N}$ (as shown in (5.9a)). For brevity in notation, we concatenate all trainable parameters of the GNU-GNNs into a vector $\boldsymbol{\omega} := [\{\Theta_i\}_{i=0}^{I}, \{\mathbf{A}_i\}_{i=0}^{I}, \{\mathbf{B}_i\}_{i=0}^{I}, \{\mathbf{b}_i^1\}_{i=0}^{I}]$. We denote the end-to-end GNU-GNNs parametric model as $\pi(\mathbf{z}; \boldsymbol{\omega})$, which, given measurements $\mathbf{z}$, predicts the voltages across all buses, represented as $\hat{\mathbf{v}} = \pi(\mathbf{z}; \boldsymbol{\omega})$. The weights ω of the GNU-GNNs can be updated through backpropagation, by specifying a suitable loss function $\ell(\mathbf{v}^*, \mathbf{v}_{I+1})$ that quantifies the agreement between the estimated voltages $\mathbf{v}_{I+1}$ produced by the GNU-GNNs and the ground-truth values $\mathbf{v}^*$. An overview of the proposed method is presented in Algorithm 1.

Algorithm 1 PSSE Solver with GNNs Priors

Training phase:

1: **Input:** raining samples $\{(\mathbf{z}^t, \mathbf{v}^{*t})\}_{t=1}^{\mathrm{T}}$

2: **Initialize:**

$\boldsymbol{\omega}^1 := [\{\Theta_i^1\}_{i=0}^{I}, \{\mathbf{A}_i^1\}_{i=0}^{I}, \{\mathbf{B}_i^1\}_{i=0}^{I}, \{\mathbf{b}_i^1\}_{i=0}^{I}]$, $\mathbf{v}_0 = 0$

3: **for** $t = 1, 2, \ldots, T$ **do**

4: Feed $\mathbf{z}^t$ and $\mathbf{v}_0$ as input into GNU-GNNs

5: **for** $i = 0, 1, \ldots, I$ **do**

6: Reshape $\mathbf{v}_i \in \mathbf{R}^{2N}$ to get $\mathbf{X}_0^i \in \mathbb{R}^{N\times 2}$

7: Feed $\mathbf{X}_0^i$ into GNNs

8: Vectorize the GNNs output $\mathbf{X}_L^i \in \mathbb{R}^{N\times 2}$ to get $\mathbf{u}_i$

9: Obtain $\mathbf{v}_{i+1} \in \mathbf{R}^{2N}$ using (5.9b)

10: **end for**

11: Obtain $\mathbf{v}_{I+1}^t$ using (5.9b)

12: Minimize the loss $\ell(\mathbf{v}^{*t}, \mathbf{v}_{I+1}^t)$ and update $\boldsymbol{\omega}^{\mathrm{T}}$

13: **end for**

14: **Output:** $\boldsymbol{\omega}^{\mathrm{T}}$

Inference phase:

1: **for** $t = T + 1, \ldots, T'$ **do**

2: Feed real-time $\mathbf{z}^t$ to the trained GNU-GNNs

3: Obtain the estimated voltage $\mathbf{v}^{\mathrm{T}}$

4: **end for**

5.3 Robust Power System State Estimation Solver

In real-time inference, our proposed GNU-GNNs, trained using past data, generates an estimate of the state $\mathbf{v}^t$ for per time slot t based on the observed measurements $\mathbf{z}^t$. However, in the presence of impulsive communication noise and possibly cyberattacks, the results obtained from our proposed GNU-GNNs (as presented in Sect. 5.2) can be significantly biased. An extension of our approach involves addressing these imperfections within the PSSE problem. Therefore, after introducing our method for incorporating prior information and training the DNNs for normal inputs, we proceed to enhance the robustness of our approach in the presence of such imperfections in this section.

To obtain robust estimators capable of handling anomalous data, classical formulations such as Huber estimation, Huber M-estimation, and Schweppe-Huber generalized M-estimation are typically based on the assumption that measurements follow ϵ-contaminated probability models; for further details, refer to (Wang et al. 2019c). Instead, the present section postulates that measured and ground-truth voltages are drawn from some nominal yet unknown distribution P_0 supported on $S = Z \times V$, that is $(\mathbf{z}, \mathbf{v}^*) \sim P_0$. Therefore, to obtain the end-to-end GNU-GNNs parametric model $\pi(\mathbf{z}; \boldsymbol{\omega})$, the trainable parameters $\boldsymbol{\omega}$ are optimized by solving $\min_{\boldsymbol{\omega}} \mathbb{E}_{P_0}[\ell(\pi(\mathbf{z}; \boldsymbol{\omega}), \mathbf{v}^*)]$. In practice, P_0 is unknown but i.i.d. training samples $\{(\mathbf{z}^t, \mathbf{v}^{*t})\}_{t=1}^{\mathrm{T}} \sim P_0$ are available. In this context, our PSSE amounts to solving for the minimizer of the empirical loss as

$$\min_{\boldsymbol{\omega}} \overline{\mathbb{E}}_{\hat{P}_0^{(T)}}[\ell(\pi(\mathbf{z}^t; \boldsymbol{\omega}), \mathbf{v}^{*t})] = \frac{1}{T} \sum_{t=1}^{T} \ell(\pi(\mathbf{z}^t; \boldsymbol{\omega}), \mathbf{v}^{*t}). \tag{5.16}$$

To address the challenges posed by uncertain and adversarial environments, we can enhance the solution of (5.16) by introducing robustness. This is achieved by optimizing over a set P of probability distributions centered around $\hat{P}_0^{(T)}$ while minimizing the expected loss under the worst-case distribution within P. This concept can be formally expressed as a distributionally robust optimization problem

$$\min_{\boldsymbol{\omega}} \sup_{P \in P} \mathbb{E}_P[\ell(\pi(\mathbf{z}; \boldsymbol{\omega}), \mathbf{v}^*)]. \tag{5.17}$$

In contrast to (5.16), the worst-case formulation in (5.17) provides a more robust performance guarantee across a diverse range of distributions within P. There exists a broad range of possible choices for the ambiguity sets P in this context. Notably, thanks to the advantageous concept of optimal transport theory (Villani et al. 2009), distributionally robust optimization techniques have become increasingly popular for enhancing the robustness of machine learning models (Bandi and Bertsimas 2014). It is this enhanced tractability that motivates the exploration of these methods in this section.

Considering two probability density functions, P and Q, defined over the support S, we denote by $\Pi(P, Q)$ the set encompassing all joint probability distributions with marginal P and Q. Furthermore, we introduce a cost function $c : Z \times Z \to [0, \infty)$, which quantifies the expense associated with transporting a unit of mass from one point, denoted as $(\mathbf{z}, \mathbf{v}^*)$ in P, to another point $(\mathbf{z}', \mathbf{v}^*)$ in Q. It's essential to note that in this context, we assume that an attacker can potentially compromise the measurements $\mathbf{z}$ but not the actual system state $\mathbf{v}^*$. The optimal transport problem between the two distributions P and Q is formally defined as (Villani et al. 2009)

$$W_c(P, Q) := \inf_{\pi \in \Pi} \mathbb{E}_{\pi}[c(\mathbf{z}, \mathbf{z}')]. \tag{5.18}$$

Intuitively, $W_c(P, Q)$ denotes the minimum cost associated with transporting all the mass from distribution P to Q. Under mild conditions over the cost function and distributions, W_c gives the well-known Wasserstein distance between P and Q; see e.g., (Sinha et al. 2017).

Introducing the distance metric W_c, we proceed to define an uncertainty set for the empirical distribution $\hat{P}_0^{(T)}$ as $P := \{P | W_c(P, \hat{P}_0^{(T)}) \leq \rho\}$, encompassing all probability distributions that deviate by at most ρ-distance from $P_0^{(T)}$. The incorporation of P into (5.17) leads to the subsequent optimization problem for distributionally robust GNU-GNNs estimation

$$\min_{\boldsymbol{\omega}} \sup_{P} \mathbb{E}_P[\ell(\pi(\mathbf{z}; \boldsymbol{\omega}), \mathbf{v}^*)], \tag{5.19a}$$

$$\text{s.t.} \quad W_c(P, \hat{P}_0^{(T)}) \leq \rho. \tag{5.19b}$$

The inner functional optimization presented in (5.19a) spans across all probability distributions P defined by the constraints in (5.19b). However, directly optimizing over these infinite-dimensional distribution functions is an intractable task. Fortunately, when dealing with continuous loss and transportation cost functions, we can determine that the optimal objective value of the inner maximization coincides with its dual optimal objective value. Moreover, the dual problem involves the optimization of only a one-dimensional variable. These two insights lead us to the decision to tackle (5.19a), (5.19b) in the dual domain. To formally derive this manageable surrogate, we invoke a result from (Blanchet and Murthy 2019).

Proposition 5.1 *Let the loss* $\ell : \boldsymbol{\omega} \times Z \times V \to [0, \infty)$*, and transportation cost* $c : Z \times Z \to [0, \infty)$ *be continuous functions. Then, for any given* $\hat{P}_0^{(T)}$*, and* $\rho > 0$*, it holds.*

$$\sup_{P \in P} \mathbb{E}_P\big[\ell(\pi(\mathbf{z}; \boldsymbol{\omega}), \mathbf{v}^*)\big] = \inf_{\gamma \geq 0} \{\overline{\mathbb{E}}_{(\mathbf{z}, \mathbf{v}^*) \sim \hat{P}_0^{(T)}} [\sup_{\zeta \in Z} \ell(\pi(\zeta; \boldsymbol{\omega}), \mathbf{v}^*) + \gamma(\rho - c(\mathbf{z}, \zeta))]\}, \tag{5.20}$$

where $P := \Big\{P | W_c(P, \hat{P}_0^{(T)}) \leq \rho\Big\}$.

Remark 5.5 Thanks to the principle of strong duality, the right-hand side in (5.20) elegantly transforms into a univariate dual reformulation of the primal problem presented on the left-hand side. This dual formulation stands in contrast to the primal one in several notable ways. Most notably, the expectation in the dual domain is computed solely over the empirical distribution $\hat{P}_0^{(T)}$, as opposed to being defined over the entire set P. Furthermore, this reformulation offers the practical advantage of obviating the necessity of determining the optimal coupling $\pi \in \Pi$ to establish P and characterizing the primal objective for all $P \in P$. This makes it an appealing and convenient approach.

Leveraging the insights from Proposition 5.1, we can reformulate the inner maximization with its dual. This leads to the following formulation for distributionally robust PSSE optimization

$$\min_{\omega} \inf_{\gamma \geq 0} \overline{\mathbb{E}}_{(\mathbf{z},\mathbf{v}^*)\sim \hat{P}_0^{(T)}}[\sup_{\zeta \in Z} \ell(\pi(\zeta; \boldsymbol{\omega}), \mathbf{v}^*) + \gamma(\rho - c(\mathbf{z}, \zeta))]. \tag{5.21}$$

Determining the optimal solution $(\boldsymbol{\omega}^*, \gamma^*)$ of (5.21) is generally a challenging task, as it entails solving the supremum separately for per observed measurement $\mathbf{z}$. This complexity poses a significant obstacle to existing minimax optimization solvers. To overcome this challenge, a common strategy is to approximate the optimal $\boldsymbol{\omega}^*$ by solving (5.21) with a preselected and fixed $\gamma > 0$, as demonstrated in previous work (Sinha et al. 2017). Notably, research has shown that for any strongly convex transportation cost function, such as $c(\mathbf{z}, \mathbf{z}') := \|\mathbf{z} - \mathbf{z}'\|_p^2$ for any $p \geq 1$, selecting a sufficiently large $\gamma > 0$ ensures that the inner maximization becomes strongly convex, enabling efficient solutions.

Note that having a fixed γ is tantamount to tuning ρ, which in turn controls the level of infused robustness. Fixing some large enough $\gamma > 0$ in (5.21), the GNU-GNNs model can be obtained by solving

$$\min_{\boldsymbol{\omega}} \overline{\mathbb{E}}_{(\mathbf{z},\mathbf{v}^*)\sim \hat{P}_0^{(T)}}[\sup_{\zeta \in Z} \psi(\boldsymbol{\omega}, \zeta; \mathbf{z}, \mathbf{v}^*)], \tag{5.22}$$

where

$$\psi(\boldsymbol{\omega}, \zeta; \mathbf{z}, \mathbf{v}^*) := \ell(\pi(\zeta; \boldsymbol{\omega}), \mathbf{v}^*) + \gamma(\rho - c(\mathbf{z}, \zeta)). \tag{5.23}$$

In a nutshell, (5.22) can be understood as first 'adversarially' perturbing the measurements $\mathbf{z}$ into ζ^* by maximizing $\psi(\cdot)$, and then it seeks a model that minimizes the empirical loss even with such perturbed inputs. This approach enhances the robustness of the model, ensuring its effectiveness in handling future data that may be compromised by adversaries. Initialized with some $\boldsymbol{\omega}^0$, and given a datum $(\mathbf{z}^t, \mathbf{v}^*)$, we form $\psi(\cdot)$ (c.f. (5.23)), and implement a single gradient ascent step for the inner maximization as follows

$$\zeta^t = \mathbf{z}^t + \eta^t \nabla_\zeta \psi(\omega^t, \zeta; \mathbf{z}^t, \mathbf{v}^{*t})|_{\zeta = \mathbf{z}^t}, \tag{5.24}$$

where $\eta^t > 0$ is the step size. Upon computing (5.24), we utilize the perturbed data ζ^t as the input, replacing the 'healthy' data $\mathbf{z}^t$, and integrate it into Algorithm 1. With the loss $\ell(\pi(\zeta^t; \boldsymbol{\omega}), \mathbf{v}^{*t})$ depending solely on the GNU-GNNs weights $\boldsymbol{\omega}$, we subsequently update the current iterate $\boldsymbol{\omega}^t$ once more through back propagation.

5.4 Numerical Tests

This section tests the estimation performance as well as robustness of our proposed methods.

5.4.1 Simulation Setup

The simulations were conducted using an NVIDIA Titan X GPU with 12 GB RAM. For our numerical experiments, we employed actual load consumption data from the 2012 Global Energy Forecasting Competition. To generate training and testing datasets, we applied the MATPOWER toolbox (Zimmerman et al. 2010) to solve the AC power flow equations using this dataset. To ensure compatibility with power demand scales, we normalized the load data and used it as input for MATPOWER. This process resulted in the creation of 1,000 pairs of measurements and corresponding ground-truth voltage values. Of these pairs, 80% were designated for training, while the remaining 20% were reserved for testing. The measurements included sending-end active power flows and voltage magnitudes, both of which were subjected to contamination by additive white Gaussian noise. We set the standard deviations of the noise added to power flows and voltage magnitudes to 0.02 and 0.01, respectively, following the methodology outlined in (Wang et al. 2019b).

A pertinent inquiry is whether the explicit integration of power network topology through a trainable regularizer yields superior performance in comparison to other competitive approaches. Furthermore, it is of considerable interest to investigate how the adoption of a distributionally robust training methodology enhances the performance of PSSE in cases involving erroneous data and potential adversarial influences. To address these questions, we conducted numerical experiments comparing our proposed approach to four baseline PSSE methods. These baselines encompassed an optimization-based method, specifically the Wirtinger-Flow Gauss–Newton algorithm as detailed in (Džafić et al. 2018), along with three data-driven methods: (i) the prox-linear network in (Zhang et al. 2019b); (ii) a 6-layer vanilla feed-forward (F)NN; and (iii) an 8-layer FNNs. The weights of these NNs were trained using the Adam optimizer to minimize the Huber loss. For the NNs-based methods (ii and iii), we trained the model weights using the Adam optimizer with a focus on minimizing the Huber loss. During training, we maintained a fixed learning rate of 10^{-3} for a duration of 500 epochs, employing a batch size of 32. To quantitatively assess the performance of each algorithm, we defined the average estimation accuracy using the following criterion.

$$\nu = \frac{1}{N} \sum_{n=1}^{N} \left\| \mathbf{v}_n - \mathbf{v}_n^* \right\|_2^2, \tag{5.25}$$

where $\mathbf{v}_n$ is the estimated voltage profile from the noisy measurements generated using $\mathbf{v}_n^*$.

5.4.2 GNU-GNNs for Regularized PSSE

In our initial experiment, we implemented the GNU-GNNs method by unrolling a total of $I = 6$ iterations of the proposed alternating minimization solver. The deep prior of the GNU-GNNs was represented using a GNNs with $K = 2$ hops and $D = 8$ hidden units, employing ReLU activation functions for each unrolled iteration. To maintain a similar number of weight parameters to the prox-linear network, the architecture of the GNU-GNNs was designed accordingly. For the initialization of the Gauss–Newton algorithm, we used a flat voltage profile. The results of this experiment, summarized in Table 5.1, provide an overview of the average performance metrics achieved by the proposed GNU-GNNs approach, the Gauss–Newton method, the prox-linear network, a 6-layer FNNs, and an 8-layer FNNs. These evaluations were conducted over a dataset comprising 200 testing samples, and the reported metrics include estimation accuracy and runtime measurements. All experiments were conducted on the IEEE 118-bus feeder.

Unquestionably, the proposed GNU-GNNs approach stands out as the top-performing method among all those subjected to testing. To be precise, the estimation accuracy achieved by the GNU-GNNs approach surpasses that of the widely adopted Gauss–Newton solver by an impressive order of magnitude, especially when applied to the intricate IEEE 118-bus feeder. Moreover, a notable advantage of the GNU-GNNs approach is its remarkable efficiency. By judiciously transferring the computational burden from inference to the training phase, the proposed approach manages to

Table 5.1 Performance of different SE approaches

Method	ν	Time
GNU-GNNs	9.42×10^{-3}	1.54×10^{-2}
Prox-linear net	1.26×10^{-2}	1.63×10^{-2}
6-layer FNNs	8.65×10^{-2}	2.89×10^{-2}
8-layer FNNs	1.93×10^{-2}	3.94×10^{-2}
Gauss–Newton	5.25×10^{-1}	13.34

reduce the runtime significantly. In fact, the running time of the GNU-GNNs approach is a striking three orders of magnitude less than that of the optimization-based alternatives when evaluated on the IEEE 118-bus feeder. This efficiency improvement represents a significant step towards making PSSE more practical and real-time.

To assess the quality of estimates generated by the proposed GNU-GNNs method, we provide a detailed analysis of voltage magnitudes and angles for the IEEE 118-bus feeder. Notably, the estimation performance of the Gauss–Newton method is notably inferior, rendering its results unsuitable for direct comparison with other alternatives. Consequently, the Gauss–Newton method has been omitted from this set of results. Figures 5.3 and 5.4 showcase the estimated voltage profiles for buses 50 and 100 during test slots 70 to 90, respectively. Furthermore, Fig. 5.5 illustrates both ground-truth and estimated voltages for the first 20 buses during test slot 80. These visualizations provide compelling evidence of the improved performance offered by our GNU-GNNs method in comparison to conventional simulated PSSE solvers.

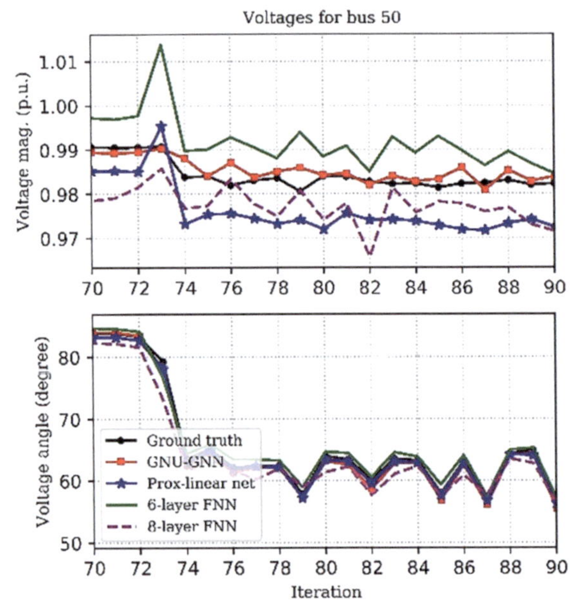

Fig. 5.3 The estimated voltage magnitudes and angles by the four schemes at bus 50 from slots 70 to 90

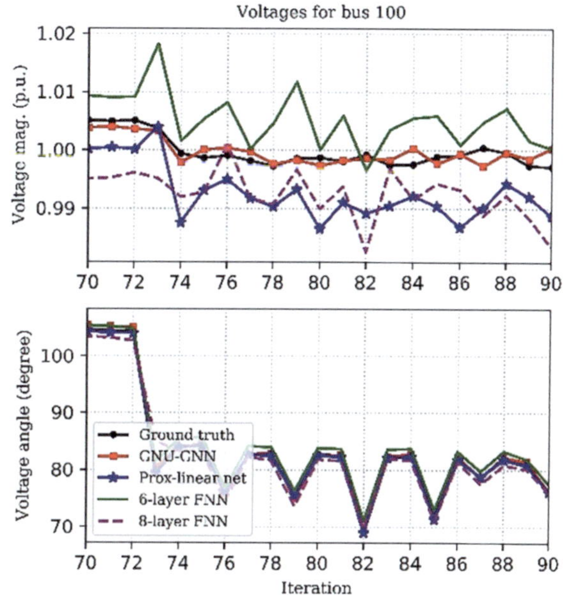

Fig. 5.4 The estimated voltage magnitudes and angles by the four schemes at bus 100 from slot 70 to 90

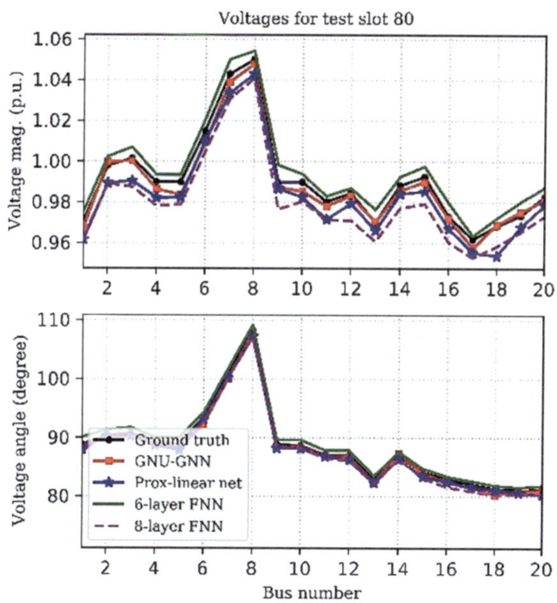

Fig. 5.5 The estimated voltages magnitudes and angles by the four schemes for the first 20 buses at slot 80

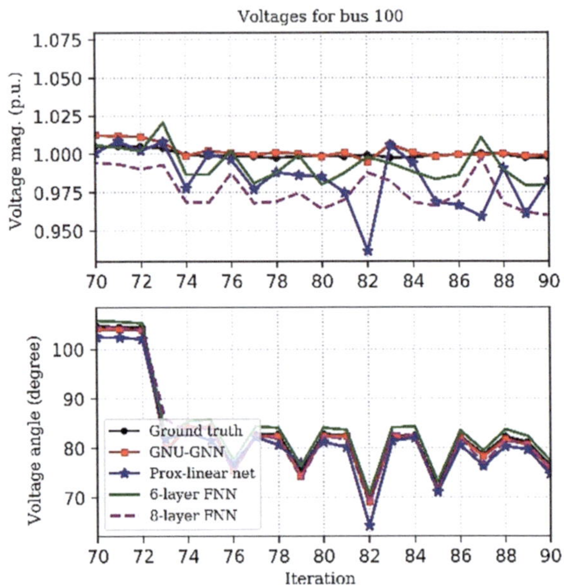

Fig. 5.6 The estimated voltage magnitudes and angles by the four schemes under distributional attacks at bus 100 from slots 70 to 90

5.4.3 *Robust PSSE*

Despite their exceptional performance in standard PSSE, DNNs may falter when confronted with unreliable or inaccurate data, a significant concern when considering their application in safety–critical power networks. In this subsection, we conduct an experiment to assess the robustness of our GNU-GNNs trained using the adversarial learning method on the IEEE 118-bus feeder. To achieve this objective, we implement a distributionally robust learning scheme that perturbs the input data for the GNU-GNNs, prox-linear network, 6-layer FNNs, and 8-layer FNNs models. Specifically, under distributional attacks, we introduce an ambiguity set $\mathcal{P}$ that comprises distributions centered around the nominal data-generating distribution P_0. While the training samples are generated according to P_0, the testing samples are drawn from a distribution $P \in \mathcal{P}$ that yields the worst empirical loss. To accomplish this, we preprocess test samples using (5.24) to generate adversarially perturbed samples. Figure 5.6 and 5.7 illustrate the estimated voltage profiles under a distributional attack with a fixed $\gamma = 0.13$ (as defined in (5.22) and (5.23)) and an ℓ_2 transportation cost. As these plots demonstrate, our proposed robust training method provides reliable performance in the face of distributional uncertainties, particularly when compared to competing alternatives.

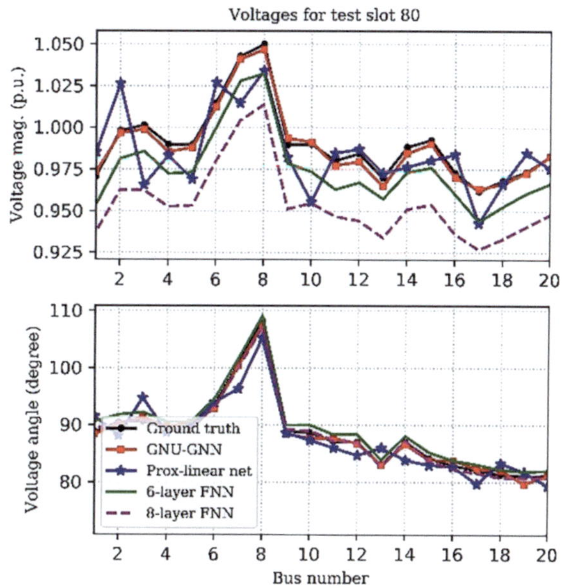

Fig. 5.7 The estimated voltage magnitudes and angles by the four schemes under distributional attacks for the first 20 buses at slot 80

Chapter 6
Stochastic Optimization via Ergodic Energy Management

The emergence of DG, along with the anticipated integration of electric vehicles and flexible loads, presents unprecedented operational challenges for distribution grids (Turitsyn et al. 2011). Indeed, many utilities in the United States are presently grappling with the complexities associated with the assimilation of residential- and commercial-scale solar generation. On the other hand, contemporary DG units are furnished with advanced smart power inverters, characterized by two-way communication and computational capabilities. Consequently, they extend novel avenues for enhanced control capabilities (Schauder 2014). In this context, recent research endeavors have concentrated on the integration of smart inverters in the energy management system (EMS) of distribution grids (Turitsyn et al. 2011; Carvalho et al. 2008; Šulc et al. 2014). Voltage regulation, employing approximate grid models, is implemented through a multi-agent framework as detailed in reference (Baran and El-Markabi 2007). Additionally, local control algorithms are developed, as expounded in reference (Šulc et al. 2014). When utilizing the comprehensive exact AC grid model, the management of reactive power falls under the domain of the non-convex optimal power flow problem, as discussed in reference (Gan et al. 2014). In addition to harnessing the reactive power capabilities of smart inverters, the implementation of active power curtailment has been recommended as a supplementary service (Tonkoski et al. 2010; Dall'Anese et al. 2014; Su et al. 2014; Von Appen et al. 2014). Notably, droop based active power curtailment has been proposed as an efficient strategy for overvoltage concerns (Tonkoski et al. 2010). Existing energy management schemes enforce inverter related and voltage regulation-related constraints at all times. Nevertheless, the operation of forthcoming grids may gain advantages from harnessing hitherto untapped system flexibilities.

To realize this mode of operation, we introduce an EEM framework. In the context of a distribution grid featuring distributed energy sources and a feed-in tariff (FIT) program, we formulate the challenges of active power curtailment and reactive power compensation as a stochastic optimization problem. We seek to enforce stringent operational constraints on average performance, while simultaneously ensuring

G. Wang et al., *Data-driven Optimization and Control for Autonomous Energy Systems*,
https://doi.org/10.1007/978-981-95-1782-4_6

that more lenient margins are consistently met. In pursuit of solutions, we develop stochastic dual subgradient solvers, drawing upon both exact and approximate grid models characterized by varying levels of complexity. Numerical tests on a real-world 56-bus distribution grid and the IEEE 123-bus test feeder relying on both grid models corroborate the advantages of the novel schemes over their deterministic alternatives.

6.1 System Modeling and Problem Formulation

Consider a distribution grid consisting of $N + 1$ buses. This grid is represented as a tree graph, denoted as $T := (N_o, L)$, where $N_o := \{0, 1, \ldots, N\}$ represents the set of nodes (buses), and $|L| = N$ signifies the cardinality of the edge (line) set $|L| = N$. It's worth noting that while distribution grids may have a structural meshed topology, they are typically operated in a radial configuration. This tree graph is rooted at the substation bus, indexed as $n = 0$, and all non-root buses are included in the set $N := \{1, \ldots, N\}$. Let's denote v_n as the squared voltage magnitude at bus n, and $p_n + jq_n$ as the complex power injection at bus n, where n belongs to the set N_o. To simplify notation, we stack nodal quantities associated with non-root buses into column vectors $\mathbf{v}$, $\mathbf{p}$, and $\mathbf{q}$.

Active and reactive power injections at bus n can be decomposed into their generation and consumption constituents, denoted as $p_n := p_n^g - p_n^c$ and $q_n := q_n^g - q_n^c$, respectively. In the case of a purely load bus, the consumption components (p_n^c, q_n^c) are typically related by a constant power factor, with $p_n^g = q_n^g = 0$. On the other hand, a DG bus not only features nonnegative components (p_n^c, q_n^c) but also contributes renewable generation $p_n^g \geq 0$ and reactive power support q_n^g. From this point forward, we will consider all buses as constant power buses. For buses equipped solely with a shunt capacitor, we have $p_n^g = p_n^c = q_n^c = 0$ and $q_n^g > 0$. The generation and consumption components are organized into vectors $\mathbf{p}^c$, $\mathbf{p}^g$, $\mathbf{q}^c$, and $\mathbf{q}^g$ accordingly.

The energy management controller is centrally operated by the utility, responsible for communicating set-points to DG units. While coordinating the control of power inverters and utility-owned voltage regulating devices would only improve performance, it is a complex task and falls outside the scope of this discussion. For a dynamic programming approach addressing this coordination, please refer to (Farivar et al. 2011). In the proposed case, grid operation is segmented into short control periods, denoted by the index t. The length of these periods is contingent on the variability of active and reactive power consumption, as well as the availability of data predictions. Given that power inverters offer a continuously adjustable control variable, with millisecond-level responsiveness, transients in their operation have been observed to be negligible, as documented in (Liu and Bebic 2008). This stands in contrast to conventional voltage regulating equipment, which may introduce switching transients. Without loss of generality, we will assume a control interval of 30-s for the remainder of this discussion.

In time period t, the grid operator has the capability to either purchase or supply energy, denoted as $p_{0,t}$, to or from the primary grid via the substation bus, utilizing the real-time market. The associated price for this energy exchange is denoted as $\pi_{0,t}$, and it is assumed to remain positive at all times. Notably, in cases where the real-time market operates on a 5-min basis, the price $\pi_{0,t}$ remains constant for ten consecutive control periods. Within the distribution grid, customers who possess renewable energy sources, such as PV panels or wind micro-turbines, can participate in a FIT program, as described in (Couture and Cory 2009, May 2013). Under this program, any surplus renewable energy produced by a customer may be purchased by the utility company at the FIT price, denoted as $\pi_{f,t}$. While FIT prices are typically adjusted on a monthly basis, we consider time-varying prices $\pi_{f,t}$ for the sake of generality, with the assumption that these prices are positive. Energy consumption, whether from FIT customers or regular consumers, is billed at a retail price, denoted as $\pi_{r,t}$. The energy cost incurred by customer n during period t is calculated as the product of

$$\pi_{r,t}[p^g_{n,t} - p^c_{n,t}]_- - \pi_{f,t}[p^g_{n,t} - p^c_{n,t}]_+, \tag{6.1}$$

times the duration of the control period, where the operators $[a]_+ := \max\{a, 0\}$ and $[a]_- := \max\{0, -a\}$ (Zhang et al. 2013). Evidently, in per slot t, at most one of the two summands in Eq. (6.1) exhibits a nonzero value.

From the utility side, the energy cost for time slot t is

$$\pi_{0,t}p_{0,t}(\mathbf{p}^g_t, \mathbf{q}^g_t) + \pi_{f,t}\mathbf{1}^{\mathrm{T}}[\mathbf{p}^g_t - \mathbf{p}^c_t]_+ - \pi_{r,t}\mathbf{1}^{\mathrm{T}}[\mathbf{p}^g_t - \mathbf{p}^c_t]_-, \tag{6.2}$$

multiplied by the slot duration, where $[\mathbf{a}]_+$ and $[\mathbf{a}]_-$ for vectors are applied entry-wise now Heed that the energy exchange with the main grid $p_{0,t}(\mathbf{p}^g_t, \mathbf{q}^g_t)$ depends on the internal consumption and generation, and the associated power losses on distribution lines. Hence, the energy exchange, denoted as $p_{0,t}$, can be conceptualized as a function primarily dependent on the control variables $(\mathbf{p}^g_t, \mathbf{q}^g_t)$. Its reliance on the variables $(\mathbf{p}^c_t, \mathbf{q}^c_t)$ and the associated grid power losses is implicitly denoted by the subscript t.

If the energy management scheme were designed to minimize the utility's cost as defined in (6.2), it would inherently aim to minimize local generation to the greatest extent possible. To illustrate this point, let's consider a node n where, at time t, the electricity demand exceeds the installed solar capacity. Consequently, $-[p^g_{n,t} - p^c_{n,t}]_- = p^g_{n,t} - p^c_{n,t} < 0$. Then, the utility EMS would command $p^g_{n,t} = 0$ unless there is an under-voltage condition. However, this policy goes against the core goal of an FIT program. The main aim of the FIT program is to restrict the reduction of renewable power generation only when a customer has excess energy, and this excess energy cannot be purchased due to financial constraints or voltage regulation limitations. To align with the FIT terms correctly, the utility avoids reducing renewable generation when the net injection is negative. Consequently, the cost function that needs to be minimized by the energy management scheme becomes $\pi_{0,t}p_{0,t}(\mathbf{p}^g_t, \mathbf{q}^g_t) + \pi_{f,t}\mathbf{1}^{\mathrm{T}}[\mathbf{p}^g_t - \mathbf{p}^c_t]_+$ rather than that in (6.2).

Operation of the energy management scheme is outlined in the following. Prior to control period t, the EMS gathers predictions for prices $(\pi_{0,t}, \pi_{f,t})$, loads $(\mathbf{p}_t^c, \mathbf{q}_t^c)$, and the maximum available renewable generation $\overline{\mathbf{p}}_t^g$. At every period t, buses are partitioned to those having a renewable energy surplus comprising the set

$$N_t := \{n \in N : \overline{p}_{n,t}^g \geq p_{n,t}^c\}, \tag{6.3}$$

and to those buses belonging to the complementary set of N_t denoted by $\overline{N}_t$. In order to concurrently address active power curtailment and reactive power management, the EMS can systematically solve the ensuing problem for per time interval t

$$J_{1,t}^* := \min_{\mathbf{p}_t^g, \mathbf{q}_t^g} \pi_{0,t} p_{0,t}(\mathbf{p}_t^g, \mathbf{q}_t^g) + \pi_{f,t} \mathbf{1}^{\mathrm{T}} [\mathbf{p}_t^g - \mathbf{p}_t^c]_+, \tag{6.4a}$$

$$\text{s.to} \quad 0 \leq p_{n,t}^g \leq \overline{p}_{n,t}^g, \forall n \in N_t, \tag{6.4b}$$

$$p_{n,t}^g = \overline{p}_{n,t}^g, \forall n \in \overline{N}_t, \tag{6.4c}$$

$$(p_{n,t}^g)^2 + (q_{n,t}^g)^2 \leq s_n^2, \tag{6.4d}$$

$$v_l \leq v_{n,t}(\mathbf{p}_t^g, \mathbf{q}_t^g) \leq v_u, \forall n. \tag{6.4e}$$

The power injections $\{(p_{n,t}^g, q_{n,t}^g)\}_n$ are subject to constraints defined by (6.4b)–(6.4e). Constraints (6.4b)–(6.4d) apply to each individual bus n, while the voltage constraints in (6.4e) link power injections across the entire grid. Specifically, in (6.4b), the term $\overline{p}_{n,t}^g - p_{n,t}^g$ represents the active power curtailed by all inverters with surplus renewable energy at time t (when $n \in N_t$). Constraint (6.4d) is a result of the maximum apparent power capacity, known as the nameplate rating, of inverter n. Constraint (6.4e) ensures that squared voltage magnitudes remain within the prescribed interval $V := [v_l, v_u]$ at all nodes. The exact functional relationships depend on the specific grid model, as discussed in detail in Sect. 6.4. For the sake of simplicity, we have omitted apparent power flow limits on distribution lines in this explanation. However, you can easily include these constraints using the grid model described in Sect. 6.4.2. It's important to note that if the utility is willing to accept any surplus energy as long as grid constraints are met, you can achieve this by setting FIT prices $(\pi_{f,t})$ to zero for all time periods in (6.4a)–(6.4e).

Problem (6.4a)–(6.4e) ensures the satisfaction of all power and voltage constraints at all times. However, as future distribution grids evolve, they will offer opportunities to exploit flexibility that can lead to reduced operational costs and improved integration of renewable energy sources. Two possible sources of flexibility are the overloading capacity of smart inverters and the regulation ranges for voltage levels. Concerning the overloading capability, grid-tied power inverters have the potential to exceed their apparent power capacity for short durations. In practice, power electronics are often designed to operate at even 1.2–1.5 times higher than their nameplate rating, as evidenced in empirical studies (Blackburn and Domin 2006). Regarding voltage regulation, instead of enforcing squared voltage magnitudes to lie in V at every t, it may suffice for their time-averages to remain in V, allowing for wider instantaneous variations. For example, according to standard EN 50160, voltages are required to stay within V for 95% of any 10-min sample (Standard 2004). It's important to note that problem (6.4a)–(6.4e) relies on predictions $\left(\mathbf{p}_t^c, \mathbf{q}_t^c, \overline{\mathbf{p}}_t^g\right)$ and prices $\left(\pi_{0,t}, \pi_{f,t}\right)$. Therefore, it can be considered optimal only when load demand, renewable generation, and prices are perfectly known. In practice though $\left(\mathbf{p}_t^c, \mathbf{q}_t^c, \overline{\mathbf{p}}_t^g, \pi_{0,t}, \pi_{f,t}\right)$ involve uncertainties (e.g. measurement noise, time-delay, and system variability) rendering the solution of (6.4a)–(6.4e) hardly optimal.

To harness operational flexibilities and address uncertainties, we transition from the deterministic formulation of energy management in Eqs. (6.4a)–(6.4e) to a stochastic approach. To facilitate this shift, we treat the dynamic parameters of the problem, denoted as $\left\{\mathbf{p}_t^c, \mathbf{q}_t^c, \overline{\mathbf{p}}_t^g, \pi_{0,t}, \pi_{f,t}\right\}$, as stochastic processes (Carta et al. 2009; Nogales and Conejo 2006; Giannakis et al. 2013). To capture ensemble averages through time averages, we assume that these stochastic processes are stationary and ergodic, although not necessarily independent across time (Conejo et al. 2010; Gatsis and Marques 2014). A stochastic process is called ergodic if you can understand its characteristics, like the average, from just one observation. This new energy management approach solves the following stochastic optimization problem

$$J_2^* := \min_{\{\mathbf{p}_t^g, \mathbf{q}_t^g\}_t} \mathbb{E}\left[\pi_{0,t} p_{0,t}(\mathbf{p}_t^g, \mathbf{q}_t^g) + \pi_{f,t} \mathbf{1}^{\mathrm{T}}[\mathbf{p}_t^g - \mathbf{p}_t^c]_+\right], \tag{6.5a}$$

$$\text{s.to} \quad 0 \le p_{n,t}^g \le \overline{p}_{n,t}^g, \forall n \in N_t, \tag{6.5b}$$

$$p_{n,t}^g = \overline{p}_{n,t}^g, \forall n \in \overline{N}_t, \tag{6.5c}$$

$$(p_{n,t}^g)^2 + (q_{n,t}^g)^2 \le \overline{s}_n^2, \forall n, \tag{6.5d}$$

$$\underline{v}_l \le v_{n,t}(\mathbf{p}_t^g, \mathbf{q}_t^g) \le \overline{v}_u, \forall n, \tag{6.5e}$$

$$\mathbb{E}\left[(p_{n,t}^g)^2 + (q_{n,t}^g)^2\right] \le s_n^2, \forall n, \tag{6.5f}$$

$$v_l \le \mathbb{E}\left[v_{n,t}(\mathbf{p}_t^g, \mathbf{q}_t^g)\right] \le v_u, \forall n, \tag{6.5g}$$

where the optimization variables consist of $(\mathbf{p}_t^g, \mathbf{q}_t^g)$ for all periods t, and the expectations are taken over the joint distribution of $\left(\mathbf{p}_t^c, \mathbf{q}_t^c, \overline{\mathbf{p}}_t^g, \pi_{0,t}, \pi_{f,t}\right)$ across all periods t. Constraint (6.5f) guarantees that the average apparent power complies with the nameplate inverter capacity for all buses. Meanwhile, constraint (6.5d) imposes a hard limit s_n ($s_n \geq s_n$) on the instantaneous apparent power for all nodes n. Likewise, constraint (6.5g) ensures that the average squared voltage magnitudes remain within the prescribed region V. Additionally, constraint (6.5e) guarantees that the instantaneous values of voltage magnitudes fall within an extended region denoted as $V' := \left[\underline{v}_l, \overline{v}_u\right]$, where $V \subseteq V'$.

Let's compare the solutions obtained from (6.5a)–(6.5g) with the minimizers obtained from (6.4a)–(6.4e) for every time t. It's important to note that constraint (6.4d) implies constraints (6.5d) and (6.5f), but not the other way around. Similarly, constraints (6.4e) guarantees (6.5e) and (6.5g), but not vice versa. Consequently, the stochastic scheme presented in (6.5a)–(6.5g) can be seen as a relaxation of the deterministic problem (6.4a)–(6.4e) that is solved over time t. Consequently, the minimizers of (6.5a)–(6.5g) have the potential to yield a lower average operational cost, expressed as $J_2^* \leq \mathbb{E}[J_{1,t}^*]$, where the expectation is taken over time t.

The stochastic problem presented in (6.5a)–(6.5g) involves an infinite number of variables, namely $\{\mathbf{p}_t^g, \mathbf{q}_t^g\}_t$. At time t, nodal power injections must adhere to the instantaneous constraints (6.5b)–(6.5e). Moreover, the infinite variables are interconnected over time through the objective function and the average constraints (6.5f) and (6.5g), posing significant challenges to solving (6.5a)–(6.5e). In the following section, we explore a stochastic optimization approach to address (6.5a)–(6.5e).

6.2 Ergodic Energy Management

The main goal of EEM is to create algorithms for observing predictions sequentially, such as $\left(\mathbf{p}_t^c, \mathbf{q}_t^c, \overline{\mathbf{p}}_t^g, \pi_{0,t}, \pi_{f,t}\right)$, and efficiently solving the stochastic problem in (6.5a)–(6.5e). EEM is inspired by principles from resource allocation in wireless communication networks. Due to uncertainties and variations in signal propagation, the focus shifts from optimizing instant system behavior to its average performance (Gatsis et al. 2010; Ribeiro 2010). A underlying assumption in EEM is that only the realisations of these stochastic processes are accessible, while their joint probability density functions typically remain unknown.

Given that the optimization variables, hereafter collectively denoted as $\mathbf{x} := \left(\left\{\mathbf{p}_t^g, \mathbf{q}_t^g\right\}_t\right)$, exhibit coupling through expectations, constraints (6.5f) and (6.5g) undergo dualization. Let $\boldsymbol{\nu} \in \mathbb{R}_+^N$, $\underline{\boldsymbol{\xi}} \in \mathbb{R}_+^N$, and $\overline{\boldsymbol{\xi}} \in \mathbb{R}_+^N$ represent the dual variables corresponding to (6.5f) and the lower and upper voltage bounds in (6.5g), respectively. All other constraints remain explicit. With these definitions, the Lagrangian function for (6.5a)–(6.5e) is readily expressed as

$$\begin{aligned}\mathcal{L}\Big(\mathbf{x};\boldsymbol{\nu},\underline{\boldsymbol{\xi}},\overline{\boldsymbol{\xi}}\Big) &:= \mathbb{E}\big[\pi_{0,t}p_{0,t}(\mathbf{p}_t^g,\mathbf{q}_t^g)+\pi_{f,t}\mathbf{1}^{\mathrm{T}}[\mathbf{p}_t^g-\mathbf{p}_t^c]_+\big]\\ &+\sum_{n=1}^{N}\nu_n\big\{\mathbb{E}\big[(p_{n,t}^g)^2+(q_{n,t}^g)^2\big]-s_n^2\big\}+\sum_{n=1}^{N}\underline{\xi}_n\big\{v_l-\mathbb{E}\big[v_{n,t}(\mathbf{p}_t^g,\mathbf{q}_t^g)\big]\big\}\\ &+\sum_{n=1}^{N}\overline{\xi}_n\big\{\mathbb{E}\big[v_{n,t}(\mathbf{p}_t^g,\mathbf{q}_t^g)\big]-v_u\big\}.\end{aligned}\tag{6.6}$$

The dual function for problem (6.5a)–(6.5e) is defined as the minimum of the Lagrangian function over all primal variables. Because of the linearity of the expectation operator, the minimization and expectation operators can be interchanged. Rearranging terms, the dual function is thus expressed as

$$g(\boldsymbol{\nu},\underline{\boldsymbol{\xi}},\overline{\boldsymbol{\xi}}) := \mathbb{E}\Big[g_t(\boldsymbol{\nu},\underline{\boldsymbol{\xi}},\overline{\boldsymbol{\xi}})\Big]-\sum_{n=1}^{N}\Big(\nu_n s_n^2-\underline{\xi}_n v_l+\overline{\xi}_n v_u\Big),$$

where functions $g_t(\boldsymbol{\nu},\underline{\boldsymbol{\xi}},\overline{\boldsymbol{\xi}})$ are defined as

$$\begin{aligned}g_t(\boldsymbol{\nu},\underline{\boldsymbol{\xi}},\overline{\boldsymbol{\xi}}) := &\min_{(\mathbf{p}_t^g,\mathbf{q}_t^g)\in\Omega_t}\{\pi_{0,t}p_{0,t}(\mathbf{p}_t^g,\mathbf{q}_t^g)+\pi_{f,t}\mathbf{1}^{\mathrm{T}}[\mathbf{p}_t^g-\mathbf{p}_t^c]_+\\ &+\sum_{n=1}^{N}\nu_n\big[(p_{n,t}^g)^2+(q_{n,t}^g)^2\big]+\sum_{n=1}^{N}(\overline{\xi}_n-\underline{\xi}_n)v_{n,t}(\mathbf{p}_t^g,\mathbf{q}_t^g)\},\end{aligned}\tag{6.7}$$

and the feasible set Ω_t is defined by the instantaneous constraints presented in (6.5a)–(6.5e), and it can be expressed as follows

$$\Omega_t := \{(\mathbf{p}_t^g,\mathbf{q}_t^g)\quad\text{satisfying (6.5b)}-\text{(6.5e)}.\tag{6.8}$$

The dual problem is obtained by maximizing the dual function over the dual variables, that is

$$g(\boldsymbol{\nu}^*,\underline{\boldsymbol{\xi}}^*,\overline{\boldsymbol{\xi}}^*) := \max_{\boldsymbol{\nu},\underline{\boldsymbol{\xi}},\overline{\boldsymbol{\xi}}\geq\mathbf{0}} g(\boldsymbol{\nu},\underline{\boldsymbol{\xi}},\overline{\boldsymbol{\xi}}).\tag{6.9}$$

Evaluating $g(\boldsymbol{\nu},\underline{\boldsymbol{\xi}},\overline{\boldsymbol{\xi}})$ necessitates solving an infinite number of problems of the form in (6.7) and subsequently averaging the optimal costs over the joint probability density function of $\{\mathbf{p}_t^c,\mathbf{q}_t^c,\overline{\mathbf{p}}_t^g,\pi_{0,t},\pi_{f,t}\}$. Even if the joint probability density function were available, calculating the expectations would remain a non-trivial task. Consequently, even evaluating the dual function becomes a challenging endeavor. To maximize the dual function in a feasible manner, a stochastic optimization solver is proposed next.

6.3 Stochastic Approximation Solver

The problem at hand is addressed employing a stochastic dual subgradient method, as detailed in references (Gatsis et al. 2010; Ribeiro 2010; Redondo and Conejo 1999). To maximize $g(\boldsymbol{\nu}, \underline{\boldsymbol{\xi}}, \overline{\boldsymbol{\xi}})$, the Lagrange multipliers are iteratively updated through projected subgradient iterations, employing a chosen step size $\mu > 0$, according to the formula

$$\boldsymbol{\nu}_t := \left[\boldsymbol{\nu}_{t-1} + \mu \boldsymbol{\delta}_{\nu,t}\right]_+, \tag{6.10a}$$

$$\underline{\boldsymbol{\xi}}_t := \left[\underline{\boldsymbol{\xi}}_{t-1} + \mu \boldsymbol{\delta}_{\underline{\xi},t}\right]_+, \tag{6.10b}$$

$$\overline{\boldsymbol{\xi}}_t := \left[\overline{\boldsymbol{\xi}}_{t-1} + \mu \boldsymbol{\delta}_{\overline{\xi},t}\right]_+, \tag{6.10c}$$

where the vector $\boldsymbol{\delta}_t := [\boldsymbol{\delta}_{\nu,t}^{\mathrm{T}}\ \boldsymbol{\delta}_{\underline{\xi},t}^{\mathrm{T}}\ \boldsymbol{\delta}_{\overline{\xi},t}^{\mathrm{T}}]^{\mathrm{T}}$ is a subgradient of $g_t(\boldsymbol{\nu}, \underline{\boldsymbol{\xi}}, \overline{\boldsymbol{\xi}})$ evaluated at the previous iterate $(\boldsymbol{\nu}_{t-1}, \underline{\boldsymbol{\xi}}_{t-1}, \overline{\boldsymbol{\xi}}_{t-1})$. The entries of the subgradient vector, denoted by $[\boldsymbol{\delta}_t]_n$, can be found as

$$[\boldsymbol{\delta}_{\nu,t}]_n := (\hat{p}_{n,t}^g)^2 + (\hat{q}_{n,t}^g)^2 - s_n^2, \tag{6.11a}$$

$$[\boldsymbol{\delta}_{\underline{\xi},t}]_n := v_l - v_{n,t}(\hat{\mathbf{p}}_t^g, \hat{\mathbf{q}}_t^g), \tag{6.11b}$$

$$[\boldsymbol{\delta}_{\overline{\xi},t}]_n := v_{n,t}(\hat{\mathbf{p}}_t^g, \hat{\mathbf{q}}_t^g) - v_u, \tag{6.11c}$$

for all n, where $(\hat{\mathbf{p}}_t^g, \hat{\mathbf{q}}_t^g)$ are the minimizers of the problem in (6.7) for $g_t(\boldsymbol{\nu}_{t-1}, \underline{\boldsymbol{\xi}}_{t-1}, \overline{\boldsymbol{\xi}}_{t-1})$. Note that the Lagrange multipliers are updated at every control interval.

Algorithm 1 Ergodic Energy Management

Input operational limits $\{s_n, \overline{s}_n\}_{n\in N}$, (v_l, v_u), $(\underline{v}_l, \overline{v}_u)$, and step size $\mu > 0$
Dual variables $\boldsymbol{\nu}_0$, $\underline{\boldsymbol{\xi}}_0$, and $\overline{\boldsymbol{\xi}}_0$ are initialized to zero
For $t = 1, 2, \ldots$ **do**
Acquire predictions $(\mathbf{p}_t^c, \mathbf{q}_t^c, \overline{\mathbf{p}}_t^g, \pi_{0,t}, \pi_{f,t})$
Find primal variables $(\hat{\mathbf{p}}_t^g, \hat{\mathbf{q}}_t^g)$ as the minimizers of $g_t(\boldsymbol{\nu}_{t-1}, \underline{\boldsymbol{\xi}}_{t-1}, \overline{\boldsymbol{\xi}}_{t-1})$ by solving (5.17) or (5.20)
Update dual variables $(\boldsymbol{\nu}_t, \underline{\boldsymbol{\xi}}_t, \overline{\boldsymbol{\xi}}_t)$ using (6.10a), (6.10b) and (6.10c)
Communicate setpoints $(\hat{\mathbf{p}}_t^g, \hat{\mathbf{q}}_t^g)$ to DG
End for

Algorithm 1 provides an overview of the EEM algorithm. The process begins with Step 1, where operational limits are configured, and an appropriate step size is selected. Step 2 initializes the Lagrange multipliers to zero. The EEM then proceeds through a cyclic process consisting of four key steps. In Step 4, predictions for the random variables involved are collected. In cases where more advanced prediction methods are unavailable, the most recent observed or metered values can serve as predictions for the upcoming control period of interest. Step 5 focuses on determining the optimal primal variables by solving (6.7) with the current Lagrange multipliers. Step 6 updates the Lagrange multipliers using the dual subgradient rule presented in (6.10a), (6.10b) and (6.10c). Finally, Step 7 involves communicating the calculated setpoints $(\hat{\mathbf{p}}_t^g, \hat{\mathbf{q}}_t^g)$ to the DG and applying them to the grid. It's essential to highlight that the proposed EEM scheme does not rely on specific distributional knowledge of the input data $\left(\mathbf{p}_t^c, \mathbf{q}_t^c, \overline{\mathbf{p}}_t^g, \pi_{0,t}, \pi_{f,t}\right)$. Furthermore, while the primary focus is on utility cost minimization, the framework can be adapted to address other energy management objectives, such as voltage regulation and conservation voltage reduction.

In the context of convergence analysis, it is essential to acknowledge that all primal and dual iterates are inherently influenced by the specific realizations $(\mathbf{p}_t^c, \mathbf{q}_t^c, \overline{\mathbf{p}}_t^g, \pi_{0,t}, \pi_{f,t})$. Consequently, any assertions regarding convergence are inherently probabilistic in nature. Using the definition of $\boldsymbol{\delta}_t$, it is easy to show that there exists a finite H such that $\mathbb{E}[\|\boldsymbol{\delta}_t\|_2^2 | \boldsymbol{\nu}_{t-1}, \underline{\boldsymbol{\xi}}_{t-1}, \overline{\boldsymbol{\xi}}_{t-1}] \leq H^2$ for all t, i.e., the subgradient $\boldsymbol{\delta}_t$ is bounded at all times. In particular, it holds that $[\boldsymbol{\delta}_{\nu,t}]_n^2 \leq \overline{s}_n^2$, while $[\boldsymbol{\delta}_{\underline{\xi},t}]_n^2$ and $[\boldsymbol{\delta}_{\overline{\xi},t}]_n^2$ are both upper bounded by $(\overline{v}^u - \underline{v}_l)^2$. Thus, the bound H can be selected as

$$H := \sum_{n=1}^{N} \left[\overline{s}_n^2 + 2(\overline{v}^u - \underline{v}_l)^2\right]. \tag{6.12}$$

Adopting Ribeiro (2010), the following result characterizes the almost sure feasibility and optimality of the EEM algorithm.

Proposition 6.1 (Ribeiro 2010). *For the sequences $\{\hat{\mathbf{p}}_t^g, \hat{\mathbf{q}}_t^g\}_t$ generated by the algorithm in Algorithm 1, the next hold with probability 1 for all $n \in N$.*

$$\lim_{t\to\infty} \frac{1}{t} \sum_{\tau=1}^{t} \left[(\hat{p}_{n,\tau}^g)^2 + (\hat{q}_{n,\tau}^g)^2\right] \leq s_n^2, \tag{6.13a}$$

$$v_l \leq \lim_{t\to\infty} \frac{1}{t} \sum_{\tau=1}^{t} v_{n,\tau}(\hat{\mathbf{p}}_t^g, \hat{\mathbf{q}}_t^g) \leq v_u. \tag{6.13b}$$

Furthermore, the incurred operational costs satisfy

$$\lim_{t\to\infty} \frac{1}{t} \sum_{\tau=1}^{t} [\pi_{0,\tau} p_{0,\tau}(\hat{\mathbf{p}}_t^g, \hat{\mathbf{q}}_t^g) + \pi_{f,\tau} \mathbf{1}^{\mathrm{T}} [\hat{\mathbf{p}}_t^g - \mathbf{p}_\tau^c]_+] - J_2^* \leq \frac{\mu H^2}{2},$$

almost surely for H as in (6.12).

The proof of Proposition 6.1 is available in (Ribeiro 2010). This proposition establishes that the sequences of primal variables, $\{\hat{\mathbf{p}}_t^g, \hat{\mathbf{q}}_t^g\}_t$, are almost surely feasible, ensuring the satisfaction of constraints (6.5f) and (6.5g) with high probability. Additionally, the ergodic limit of the objective function is guaranteed to be within at most $\mu H^2/2$ of the optimal value J_2^* [cf. (6.5a)–(6.5e)]. Remarkably, these assertions remain valid even in cases where the stochastic processes exhibit correlations across time (Ribeiro 2010). Although stochastic processes have been assumed to be ergodic for the theoretical claims to hold, the numerical tests in Sect. 6.5 using real data show the efficacy of the scheme even with non-ergodic data.

The EEM problem in (6.5a)–(6.5e) and its stochastic approximation solver presented in Algorithm 1 involve the consideration of two important factors: power losses denoted as $p_{0,t}(\mathbf{p}_t^g, \mathbf{q}_t^g)$ and the squared voltage magnitudes represented as $\{v_{n,t}(\mathbf{p}_t^g, \mathbf{q}_t^g)\}_n$. Up to this point, both of these variables have been expressed as functions of the control variables, $(\mathbf{p}_t^g, \mathbf{q}_t^g)$. In this context, the EEM scheme provides a versatile framework that can accommodate various power system models. To effectively execute Step 5 of the algorithm outlined in Algorithm 1, it becomes crucial to define the specific mathematical forms of $p_{0,t}(\mathbf{p}_t^g, \mathbf{q}_t^g)$ and $\{v_{n,t}(\mathbf{p}_t^g, \mathbf{q}_t^g)\}_n$. As a turnkey application of EEM, the ensuing section focuses on radial single-phase distribution grids using two power flow models with different accuracy-complexity trade-offs.

6.4 Grid Modeling and Algorithms

In this section, we will define the functions $p_{0,t}(\mathbf{p}_t^g, \mathbf{q}_t^g)$ and $\{v_{n,t}(\mathbf{p}_t^g, \mathbf{q}_t^g)\}_n$ by considering two different approaches: one based on an exact full AC grid model and the other on its linear approximation. We will then seamlessly integrate both of these cases into the EEM algorithm the choice between these models will largely depend on the available computational resources. The AC model-based EEM can be formulated as an second-order cone programming (SOCP), whereas the linear model yields a linearly constrained quadratic program. Therefore, the latter option offers an approximate yet computationally less demanding alternative to the former.

6.4.1 *Branch Flow Model-Based EEM*

Given the radial configuration of distribution grids, it is a fundamental characteristic that each non-root bus, denoted as $n \in N$, has a unique parent bus. This parent bus will be designated as α_n. Furthermore, the directed edge $(\alpha_n, n) \in L$, which represents the distribution line supplying power to bus n, will be labeled with the index n. A visual representation of this concept is illustrated in Fig. 6.1 To simplify

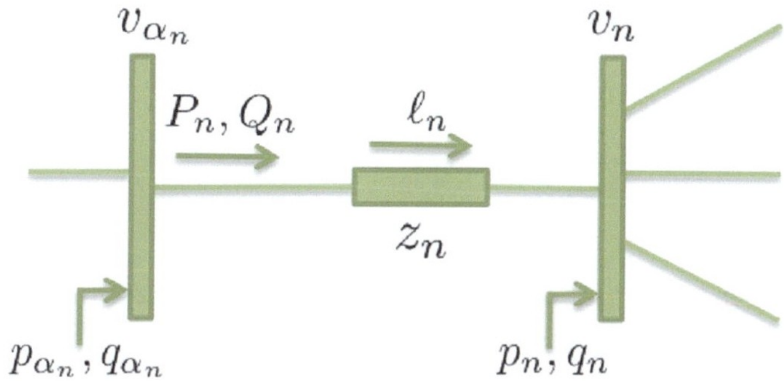

Fig. 6.1 Bus n is connected to its unique parent α_n via line n

our analysis, we can conveniently index the buses in such a way that $\alpha_n < n$ holds true for all $n \in N$, without any loss of generality.

Consider the line impedance $z_n = r_n + jx_n$ for line n, and let $\ell_{n,t}$ denote the squared current magnitude on line n at time t. If $P_{n,t} + jQ_{n,t}$ represents the complex power flow on line n observed at the sending end bus α_n during time t, we can characterize the grid using the branch-flow model (Baran and Wu 1989b)

$$p_{n,t} = \sum_{k \in C_n} P_{k,t} - (P_{n,t} - r_n \ell_{n,t}), \tag{6.14a}$$

$$q_{n,t} = \sum_{k \in C_n} Q_{k,t} - (Q_{n,t} - x_n \ell_{n,t}), \tag{6.14b}$$

$$v_{n,t} = v_{\alpha_n,t} - 2(r_n P_{n,t} + x_n Q_{n,t}) + (r_n^2 + x_n^2)\ell_{n,t}, \tag{6.14c}$$

$$\ell_{n,t} = \frac{P_{n,t}^2 + Q_{n,t}^2}{v_{\alpha_n,t}}, \tag{6.14d}$$

for all $n \in N$, where C_n is the set of the children nodes of bus n. The power injections at the substation bus are $p_{0,t} = \sum_{k \in C_0} P_{k,t}, q_{0,t} = \sum_{k \in C_0} Q_{k,t}$, and its squared voltage magnitude $v_{0,t}$ is controlled at a desirable value. Similar to $(\mathbf{p}_t, \mathbf{q}_t)$, the vectors $\mathbf{r}, \mathbf{p}_t, \mathbf{q}_t, \mathbf{v}_t$, and ℓ_t, collect the entries of $r_n, P_{n,t}, Q_{n,t}, v_{n,t}$, and $\ell_{n,t}$, accordingly.

Equations (6.14a)–(6.14c) exhibit linearity in terms of the system variables, specifically $(\mathbf{p}_t, \mathbf{q}_t, \mathbf{P}_t, \mathbf{Q}_t, \mathbf{v}_t, \ell_t)$. On the other hand, the equations presented in (6.14d) introduce quadratic terms, which in turn result in a non-convex feasible set. However, it is noteworthy that these latter equations have recently undergone a transformation, relaxing them into convex inequalities governed by hyperbolic constraints (Farivar and Low 2013)

$$P_{n,t}^2 + Q_{n,t}^2 \leq v_{\alpha_n,t} \ell_{n,t}, \forall n, \tag{6.15}$$

which can be equivalently expressed as the convex second-order cone constraints

$$\left\| \begin{bmatrix} 2P_{n,t} \\ 2Q_{n,t} \\ v_{\alpha_n,t} - \ell_{n,t} \end{bmatrix} \right\|_2 \leq v_{\alpha_n,t} + \ell_{n,t}, \forall n. \tag{6.16}$$

Equations (6.14a)–(6.14c) alongside (6.16) now define a convex feasible set. Recent research highlights the utility of this relaxed feasible set for tackling various grid optimization tasks. It is noteworthy that, under certain conditions, this relaxation has been proven to be exact, implying that the obtained minimizer satisfies (6.16) with equality; this phenomenon has been explored extensively in (Gan et al. 2014) and related studies. Henceforth, we assume that all SOCP relaxations are exact, a claim that we will substantiate through numerical validation in Sect. 6.5.

Utilizing the relationships defined by (6.14a)–(6.14c) and (6.16), we can express the active power injection at the substation bus, denoted as $p_{0,t}(\mathbf{p}_t^g, \mathbf{q}_t^g)$, as follows

$$p_{0,t}(\mathbf{p}_t^g, \mathbf{q}_t^g) = \sum_{n=1}^{N} (p_{n,t}^c - p_{n,t}^g) + \sum_{n=1}^{N} r_n \ell_{n,t} = \mathbf{1}^{\mathrm{T}}(\mathbf{p}_t^c - \mathbf{p}_t^g) + \mathbf{r}^{\mathrm{T}} \ell_t,$$

where the second part of the equation accounts for the overall power losses on distribution lines. Therefore, in this context, the primal update (Step 5 in Algorithm 1) involves solving the optimization problem

$$\min_{\mathbf{p}_t^g, \mathbf{q}_t^g, \ell_t, \mathbf{P}_t, \mathbf{Q}_t, \mathbf{v}_t} \quad \pi_{0,t}\mathbf{1}^{\mathrm{T}}(\mathbf{p}_t^c - \mathbf{p}_t^g) + \pi_{0,t}\mathbf{r}^{\mathrm{T}}\ell_t + \pi_{f,t}\mathbf{1}^{\mathrm{T}}[\mathbf{p}_t^g - \mathbf{p}_t^c]_+$$
$$+ \sum_{n=1}^{N} \nu_{n,t-1}\left[(p_{n,t}^g)^2 + (q_{n,t}^g)^2\right] + \sum_{n=1}^{N} (\overline{\xi}_{n,t-1} - \underline{\xi}_{n,t-1}) v_{n,t}, \tag{6.17}$$

s.to (6.5b)–(6.5e), (6.14a)–(6.14c), (6.16).

In addition to the original variables $(\mathbf{p}_t^g, \mathbf{q}_t^g)$, the primal update now encompasses the variables $(\mathbf{P}_t, \mathbf{Q}_t, \mathbf{v}_t, \ell_t)$ as well. Problem (6.17) can be reformulated into an SOCP. It's important to note that all instances of (6.17) solved in Sect. 6.5 were exact. Nevertheless, solving (6.17) could be computationally demanding, especially for large-scale distribution grids. This motivates our next instantiation of the EEM algorithm, which utilizes an approximate grid model.

6.4.2 Linear Distribution Flow-Based EEM

The linear distribution flow model can be derived as follows. Due to the relatively small values of the line parameters $\{r_n, x_n\}_{n\in N}$, we can neglect the last summands in (6.14a)–(6.14c), resulting in linear equations for all $n \in N$ (Baran and Wu 1989b)

$$p_{n,t} = \sum_{k \in C_n} P_{k,t} - P_{n,t}, \tag{6.18a}$$

$$q_{n,t} = \sum_{k \in C_n} Q_{k,t} - Q_{n,t}, \tag{6.18b}$$

$$v_{n,t} = v_{\alpha_n,t} - 2(r_n P_{n,t} + x_n Q_{n,t}). \tag{6.18c}$$

In this manner, we approximate squared voltage magnitudes as linear functions of $(\mathbf{p}_t, \mathbf{q}_t)$. This approximation is reasonable when squared voltage magnitudes are close to 1 per unit. Additionally, we approximate squared line current magnitudes as (Baran and Wu 1989b)

$$\ell_{n,t} = \frac{P_{n,t}^2 + Q_{n,t}^2}{v_{\alpha_n,t}} \approx P_{n,t}^2 + Q_{n,t}^2. \tag{6.19}$$

Hence, we can approximate the active power injection at the substation bus as follows

$$p_{0,t}(\mathbf{p}_t^g, \mathbf{q}_t^g) = \mathbf{1}^{\mathrm{T}}(\mathbf{p}_t^c - \mathbf{p}_t^g) + \sum_{n=1}^{N} r_n \big(P_{n,t}^2 + Q_{n,t}^2\big).$$

Expanding upon the approximate model of (6.18a), (6.18b), (6.18c) and (6.19), the primal update within the EEM algorithm (Step 5 of Algorithm 1) for time period t involves solving the following problem

$$\min_{\substack{\mathbf{p}_t^g, \mathbf{q}_t^g, \\ \mathbf{P}_t, \mathbf{Q}_t, \mathbf{v}_t}} \; \pi_{0,t} \mathbf{1}^{\mathrm{T}}(\mathbf{p}_t^c - \mathbf{p}_t^g) + \pi_{0,t} \sum_{n=1}^{N} r_n (P_{n,t}^2 + Q_{n,t}^2) + \pi_{f,t} \mathbf{1}^{\mathrm{T}}[\mathbf{p}_t^g - \mathbf{p}_t^c]_+$$

$$+ \sum_{n=1}^{N} (\overline{\xi}_{n,t-1} - \underline{\xi}_{n,t-1}) v_{n,t} + \sum_{n=1}^{N} \nu_{n,t-1} \big[(p_{n,t}^g)^2 + (q_{n,t}^g)^2\big], \tag{6.20}$$

s.to (6.5b)–(6.5e), (6.18a)–(6.18c).

Utilizing (6.18a)–(6.18b), the line flow variables $(\mathbf{P}_t, \mathbf{Q}_t)$ can be expressed as linear functions of $(\mathbf{p}_t, \mathbf{q}_t)$. Consequently, problem (6.20) can be addressed as a linearly constrained quadratic program.

6.5 Numerical Tests

The innovative methodologies underwent rigorous numerical testing on a 56-bus distribution grid sourced from Southern California Edison and the IEEE 123-bus feeder (Gan et al. 2014; Feeders 2011). You can find the line and bus data for the Southern California Edison grid in Tables 6.1 and 6.2, respectively. It's worth noting that, for all loads, a power factor of 0.8 was assumed, and you can find comprehensive details in (Gan et al. 2014). Additionally, the PV units on buses 19 and 45 were configured with a capacity of 6 MW each. During each 30-s control period, the EEM controller collected power demand data from load buses and solar generation predictions from PV units. Subsequently, it computed the active and reactive power injections by PV inverters using two distinct methods: (i) solving the DEM scheme outlined in (6.4a)–(6.4e); and (ii) employing the novel EEM algorithm outlined in Algorithm 1, which was initialized to zero.

The margins defining the squared voltage magnitudes were meticulously set as follows: $[v_l, v_u] = [0.9604, 1.0404]$ per unit and $[\underline{v}_l, \overline{v}_u] = [0.9409, 1.0609]$ per unit, with the nominal voltage $v_0 = 1$ per unit. For smart inverters, the apparent power capability was established at 1.3 times the nameplate capacity of the corresponding PV. Our evaluation criteria revolved around two pivotal metrics: the energy management cost and the instantaneous counterpart of the cost delineated in (6.5a)–(6.5e). The practical implementation of all algorithms was carried out using MATLAB and CVX, executed on a robust Intel CPU clocking in at 3.4 GHz with 32 GB of RAM (Grant et al. 2009). Impressively, each run of the full AC model and the linear approximation model-based algorithms, performed on the 56-bus grid, concluded in a mere 1.5 and 1.3 s, respectively. For the IEEE 123-bus feeder, the corresponding times increased marginally to 4.5 and 3 s. Notably, it is worth highlighting that all SOCP relaxations encountered throughout our experiments were not only feasible but also exact in nature.

To verify the almost sure optimality, our initial experiment conducted on the 56-bus grid was designed to replicate real-world cases. Synthetic load consumption and solar generation were modeled as $\mathbf{p}_t^c = \mathbf{p}^c + \epsilon_t^c$ and $\overline{\mathbf{p}}_t^g = \mathbf{p}^g + \epsilon_t^g$, respectively. For the nominal values $\mathbf{p}^c$ and $\mathbf{p}^g$, we selected values amounting to 40% of the peak demand and 80% of the maximum PV generation, respectively. To capture the inherent fluctuations in these variables, vectors ϵ_t^c and ϵ_t^g were employed, with fluctuations meticulously characterized as independent zero-mean Gaussian vectors. Their standard deviations were set at 5% of the corresponding nominal values, reflecting real-world uncertainties. Recognizing the temporal variability in FIT prices, which are typically half of the consumption prices (Von Appen et al. 2014), we assigned prices of $\pi_{0,t} = 30$/kWh and $\pi_{f,t} = 15$/kWh for all time periods t. This experimental setup allowed us to simulate conditions closely mirroring actual operational environments.

Utilizing the branch flow model, Fig. 6.2 provides an illustrative representation of the energy management cost, comparing the deterministic and EEM schemes over a single realization spanning 120 control periods. For the ergodic scheme, we

Table 6.1 Line data for the 56-bus distribution feeder (Gan et al. 2014)

From bus	To bus	r_i [Ω]	x_i [Ω]
1	2	0.160	0.388
2	3	0.824	0.315
2	4	0.144	0.349
4	5	1.026	0.421
4	6	0.741	0.466
4	7	0.528	0.468
4	20	0.138	0.334
7	8	0.358	0.314
8	9	2.032	0.798
8	10	0.502	0.441
10	11	0.372	0.327
11	12	1.431	0.999
11	13	0.429	0.377
13	14	0.671	0.257
13	15	0.457	0.401
15	16	1.008	0.385
15	17	0.153	0.134
17	18	0.971	0.722
18	19	1.885	0.721
20	21	0.251	0.096
20	23	0.225	0.542
21	22	1.818	0.695
23	24	0.127	0.542
23	25	0.284	0.687
25	26	0.171	0.414
26	27	0.414	0.386
26	32	0.205	0.495
27	28	0.210	0.196
28	29	0.395	0.369
29	30	0.248	0.232
30	31	0.279	0.260
32	33	0.263	0.073
32	34	0.071	0.171
34	35	0.625	0.273
34	36	0.510	0.209
34	38	1.062	0.406

(continued)

Table 6.1 (continued)

From bus	To bus	r_i [Ω]	x_i [Ω]
34	41	0.115	0.278
36	37	2.018	0.819
38	39	0.610	0.238
39	40	2.349	0.964
41	42	0.159	0.384
41	47	0.157	0.379
42	43	0.934	0.383
42	44	0.506	0.163
42	45	0.095	0.195
42	46	1.915	0.769
47	48	1.641	0.670
47	49	0.081	0.196
49	50	1.727	0.709
49	51	0.112	0.270
51	52	0.674	0.275
51	53	0.070	0.170
53	54	2.041	0.780
53	55	0.813	0.334
53	53	0.141	0.340

configured a step size of $\mu = 0.1$. The time-average energy management cost per time slot t is defined as $\frac{1}{t}\sum_{\tau=1}^{t}\left[\pi_{0,\tau}p_{0,\tau}(\hat{\mathbf{p}}_t^g, \hat{\mathbf{q}}_t^g) + \pi_{f,\tau}\mathbf{1}^{\mathrm{T}}[\hat{\mathbf{p}}_t^g - \mathbf{p}_\tau^c]_+\right]$. Evaluating the performance over an hour, the actual total operational cost tallies to – \$565 for the DEM and – \$589 for the EEM scheme. These results offer a comparative perspective on the operational costs of both approaches.

The second test studies the impact of the step size μ on the convergence behavior of the EEM scheme, utilizing the AC-based EEM scheme in the simulations. Step sizes ranging from $\mu = 0.1$ to $\mu = 0.3$ were explored, alongside the Deterministic Energy Management (DEM) scheme, for a total of twenty Monte Carlo system realizations spanning 60 control periods. The averaged energy management costs for each step size value are visually depicted in Fig. 6.3. The plotted curves distinctly illustrate a trend where larger step sizes result in higher energy management costs. This observation aligns with the theoretical findings regarding the optimality gap detailed in Proposition 6.1.

To rigorously assess the proposed schemes under authentic real-world conditions, we conducted two experiments employing data sourced from the Smart * project (Barker et al. 2012). In the first experiment, we employed consumption data, which entailed minute-level samples from 443 homes on April 2, 2011. Additionally, we incorporated data pertaining to the power output of 3 residential PV installations, collected at 5-s intervals over the course of August 12, 2011. Data were preprocessed

Table 6.2 Bus data for the 56-bus distribution feeder (Gan et al. 2014)

Bus No	Peak [MVA]
3	30
5	0.67
6	0.45
7	0.89
8	0.07
9	0.67
10	0.45
11	2.223
12	0.45
14	0.20
16	0.13
17	0.13
18	0.20
19	0.45
33	2.23
24	0.45
25	0.20
27	0.13
28	0.13
29	0.07
31	0.13
32	0.27
33	0.20
34	0.27
35	0.45
36	1.34
37	0.13
38	0.67
39	0.13
40	0.45
41	0.20
42	0.45
43	1.34
44	0.13
46	0.64
47	0.13
48	0.45
50	0.20

(continued)

Table 6.2 (continued)

Bus No	Peak [MVA]
Shunt capacitors	
Bus nameplate capacity	
No.	[MVA]
19	0.6
21	0.6
30	0.6
53	0.6
Base information	
$V_{\text{base}} = 12$ kV	
$S_{\text{base}} = 1$ MVA	

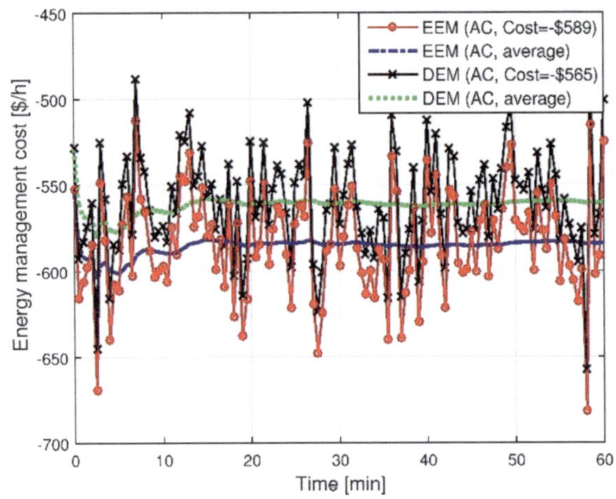

Fig. 6.2 Energy management cost using the AC model on synthetic data ($\mu = 0.1$ for EEM)

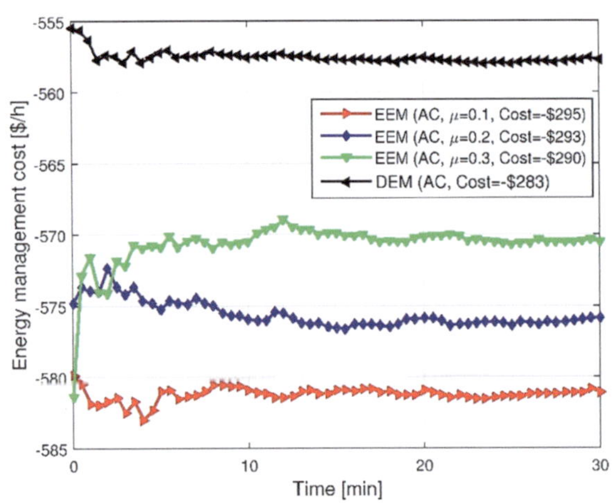

Fig. 6.3 Energy management cost averaged over 20 independent realizations using the AC model

as follows. Consumption data were first linearly interpolated to yield 30-s loads, and then averaged over every 10 homes to better resemble bus loads. Daily load curves were subsequently normalized to a maximum value of one, and mapped to different buses (Gan et al. 2014). Normalized daily load curves were multiplied by the respective nominal load values for each bus. In the case of PV systems, 5-s data was aggregated to 30-s intervals. Similarly, daily generation data was scaled to align with rated capacities.

A solitary system realization was meticulously simulated, encompassing a total of 600 consecutive 30-s control periods within the time frame of 9:30 am to 2:30 pm. We conducted this simulation for both the AC-based and the approximate model-based energy management schemes. In Fig. 6.4, we present the corresponding cost profiles for a step size of $\mu = 0.25$. Remarkably, whether employing the AC or the linear approximation model, the novel EEM scheme consistently outperforms the conventional DEM scheme, achieving a notably reduced operational cost.

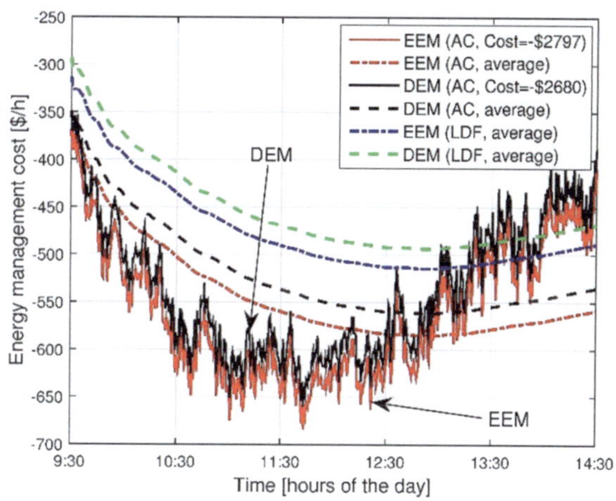

Fig. 6.4 Energy management cost using the AC and the linear distribution flow (LDF) models on real load and solar generation data (Barker et al. 2012)

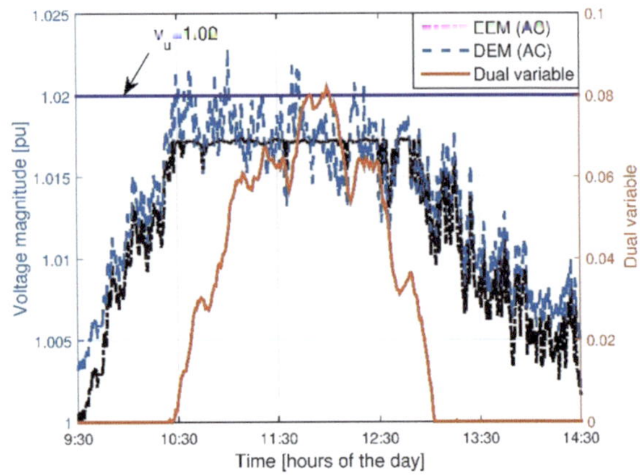

Fig. 6.5 Voltage magnitude for bus 45 and the associated dual variable $\overline{\xi}_{45,t}$ using the AC model-based schemes

Figure 6.5 provides an insightful view of the squared voltage magnitude evolution for bus 45, as well as the progression of the dual variable $\overline{\xi}_{45,t}$ associated with the stringent voltage margin constraint in (6.5g). In the case of the deterministic scheme, the voltage magnitude consistently adheres to the narrow region $[v_l, v_u] = [0.9604, 1.0404]$ throughout the entire operational horizon. Conversely, the stochastic scheme occasionally permits voltage magnitudes to temporarily exceed the upper margin of $v_u = 1.0404$. However, these over-voltage occurrences are of short duration. At approximately 10:25 am, the voltage magnitude breaches the tight voltage constraint for the first time, leading to a positive increase in the associated dual variable. As long as the voltage magnitude fluctuates above the stringent margin, the dual variable continues to rise. Around 12:20 pm, a significant shift occurs as the voltage magnitude begins to consistently fall below the upper margin. Simultaneously, the dual variable experiences a gradual decline and ultimately reaches zero for the remainder of the day.

For a comprehensive assessment of voltage regulation and active power curtailment at the grid-level, we turn our attention to Fig. 6.6. Here, we observe the grid-averaged voltage magnitudes resulting from both the DEM and EEM schemes, as well as the case without any control mechanism. In the absence of control, voltage magnitudes consistently surpass regulatory margins. In contrast, the EEM scheme offers a slightly elevated voltage profile compared to DEM, albeit at the expense of reduced operational cost. Similarly, Fig. 6.7 delves into grid-wise solar generation curtailment incurred by both the DEM and EEM schemes. It becomes evident that the DEM scheme necessitates a more pronounced curtailment of active power compared to its EEM counterpart.

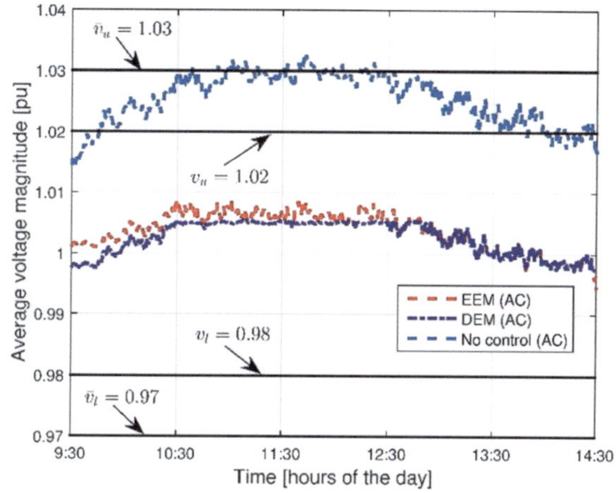

Fig. 6.6 Grid-averaged voltage magnitude using the AC model-based schemes

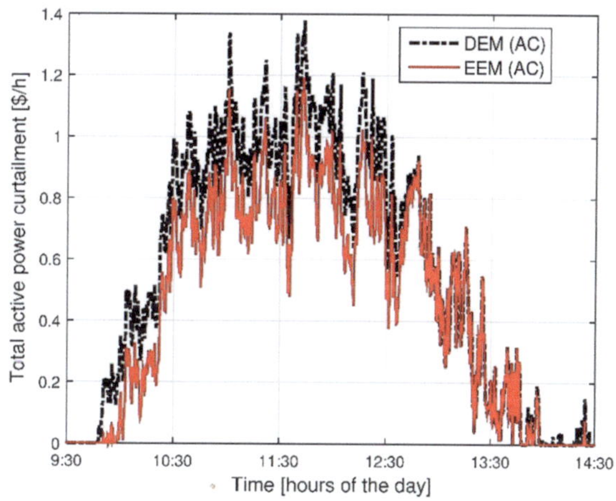

Fig. 6.7 Total active power curtailment over 9:30 am–2:30 pm

In the final test case, we examined the IEEE 123-bus feeder, as illustrated in Fig. 6.8 (Feeders 2011). To facilitate computational analysis, we converted the original multiphase system into a single-phase approximation using a heuristic approach, as detailed in (Gan et al. 2014). This transformation involved several simplifications, including the uniform distribution of loads over all phases, averaging line self-impedances across phases, neglecting mutual impedances, modeling closed switches as short circuits, and disregarding open switches. Distributed loads were substituted with two identical spot loads placed at each line's endpoints. Transformers were represented as lines with predefined impedances, and voltage regulator tap ratios were fixed at 1.08. Furthermore, a single PV system with a nameplate rating of 1.2 MW was installed at bus 114, corresponding to nearly 100 % PV penetration.

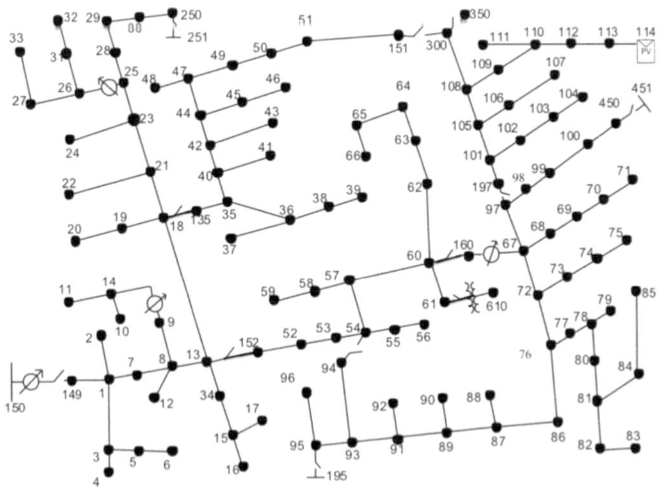

Fig. 6.8 Schematic diagram of the IEEE 123-bus test system with a PV (Feeders 2011)

With an initial voltage magnitude of 1 per unit, we set voltage regulation bounds at $[v_l, v_u] = [0.9801, 1.0201]$ per unit and $[\underline{v}_l, \overline{v}_u] = [0.9409, 1.0609]$ per unit Additionally, the inverters were allowed to operate at 110% of their nameplate rating. For this test, the linearized model was adopted, and real data for solar generation and household loads were utilized. The pricing structure consisted of $\pi_{0,t} = 30/$ kWh and $\pi_{f,t} = 15/$ kWh for per time period t. Figure 6.9 presents the cost over 600 30-s control slots spanning from 9:30 am to 2:30 pm for both the EEM and DEM schemes. A step size of $\mu = 0.001$ was selected. Throughout the simulation, the EEM approach incurred a total operational cost of $-$ 1146, while the DEM scheme resulted in a cost of $-$ 991. These results decisively demonstrate the superiority of the ergodic approach when applied to the IEEE 123-bus feeder.

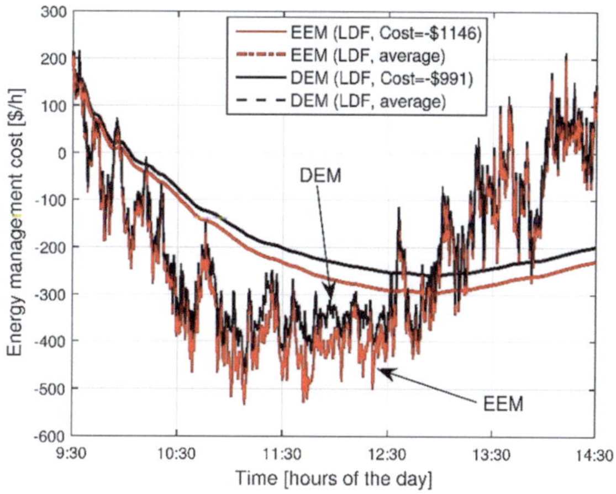

Fig. 6.9 Energy management cost evaluated on the IEEE 123-bus test system using the linearized model on real load and solar generation data

Chapter 7
Stochastic Power Control via Deep Reinforcement Learning

Frequent and substantial voltage fluctuations, attributed to the widespread adoption of electric vehicles, demand response initiatives, and the integration of renewable energy sources, present significant challenges to contemporary distribution grids. Electric utilities are currently experiencing major issues related to the unprecedented levels of load peaks as well as renewable penetration. This underscores the importance of voltage regulation, the essential task of ensuring that bus voltage magnitudes remain within optimal ranges, in modern distribution grids. The management of residential voltage levels primarily depended on utility-owned equipment such as load-tap-changing transformers, voltage regulators, and capacitor banks. Recently, a batch reinforcement learning (RL) framework utilizing linear function approximation was proposed in (Xu et al. 2019). Another characteristic of utility-owned equipment is their limited operational lifespan, necessitating daily or even monthly basis. Such configurations have been effective in traditional distribution grids without (or with low) renewable generation, and with slowly varying load.

On the other hand, smart power inverters are integral to modern DG units like PV panels and wind turbines. Equipped with computational and communication units, these inverters can be swiftly instructed to modify reactive power output in a continuous and real-time manner, responding within seconds. RL encompasses a set of methodologies designed for addressing Markovian decision processes, particularly in cases where the underlying transition mechanism is unknown (Sutton and Barto 2018). By combining deep learning with RL, specifically utilizing a DNNs to approximate the action-value function, deep reinforcement learning (DRL) has achieved performance levels in artificial agents that are comparable to human-level performance in a wide range of application domains (Mnih et al. 2015; Sadeghi et al. 2019). DRL algorithms have also demonstrated significant promise in addressing various demanding tasks related to power systems control, monitoring (Diao et al. 2019; Ernst et al. 2004; Xu et al. 2019; Zamzam et al. 2019b; Yan and Xu 2018; Lu et al. 2019), and load control (Claessens et al. 2016; Duan et al. 2017).

G. Wang et al., *Data-driven Optimization and Control for Autonomous Energy Systems*,
https://doi.org/10.1007/978-981-95-1782-4_7

In this context, by ingeniously amalgamating data-driven and physics based optimization approaches, this chapter pioneers the development of an innovative two-timescale voltage regulation scheme for the distribution grid. On a faster timescale, for instance, every second, the optimal setpoint for smart inverters is ascertained by minimizing the deviation between instantaneous bus voltage and its nominal value. This optimization is grounded in either a precise AC power flow model or its linear approximation. Conversely, on the more slower timescale, such as every hour, we harness DRL algorithms to configure parallel capacitors. This configuration minimizes the voltage deviation over the long term while considering discounted values. The effectiveness of this approach is validated through an abundance of numerical experiments conducted on an actual 47-bus distribution network and the IEEE 123-bus test feeder.

7.1 System Modeling and Problem Formulation

Consider a distribution grid comprising $N+1$ buses with the substation bus, indexed as $i=0$, serving as the root. The buses are organized into the set $N_0 := \{0\} \cup N$, and the lines into $\mathcal{L} := \{1, \ldots, \mathcal{N}\}$. For all buses $i \in N$ (excluding the substation bus), we denote their squared voltage magnitude as v_i and their complex power injection as $p_i + jq_i$. To simplify notation, we gather all nodal quantities into column vectors $\mathbf{v}$, $\mathbf{p}$, and $\mathbf{q}$.

Active power injection is split into its generation p_i^g and consumption p_i^c as $p_i := p_i^g - p_i^c$; likewise, reactive power injection is $q_i := q_i^g - q_i^c$. In distribution grids, it holds that $p_i^g = p_i^c = q_i^c = 0$ and $q_i^g > 0$ if bus i has a capacitor; while $p_i^g = q_i^g = 0$ if bus i is a purely load bus; and $p_i^c \geq 0$, $q_i^c \geq 0$, $p_i^g \geq 0$ if bus i is equipped with a DG. Let us stack generation and consumption components into vectors $\mathbf{p}^g$, $\mathbf{q}^g$, $\mathbf{p}^c$, and $\mathbf{q}^c$ accordingly. Predictions of active power consumption and solar generation $(\mathbf{p}^c, \mathbf{q}^c, \mathbf{p}^g)$ can be obtained through the hourly and real-time market (see e.g., Kekatos et al. 2014), or by running load demand (solar generation) prediction algorithms (Zhang et al. 2019b).

As previously noted, modern distribution grids encompass two categories of assets capable of participating in reactive power control: utility-owned equipment, characterized by discrete actions and a limited lifespan, and smart inverters, which can be controlled within seconds and offer a continuously-valued fashion. Historically, conventional devices have adequately addressed the need for voltage support, given the relatively gradual fluctuations in aggregate load. However, the growing presence of uncertain renewable energy sources necessitates the integration of fast-responding solutions through inverters.

In this context, our study places a central emphasis on voltage regulation, utilizing the reactive control capabilities of capacitors and inverters. It's worth noting that our framework can also accommodate various other types of reactive power control devices. To facilitate this, we divide each day into $N_{\overline{T}}$ intervals, denoted as $\tau = 1, \ldots, N_{\overline{T}}$. Within these intervals, we further break down time into N_T discrete time

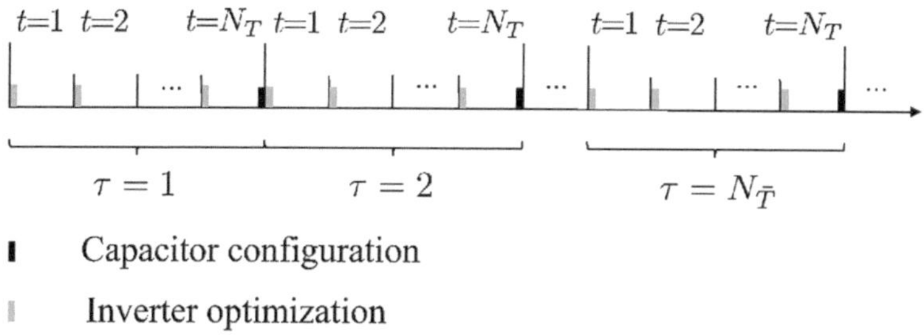

Fig. 7.1 Two-timescale partitioning of a day for joint capacitor and inverter control

slots, indexed as $t = 1, \ldots, N_T$, as shown in Fig. 7.1. To match the slow load variations, the on–off decisions of capacitors are made (at the end of) every interval τ, which can be chosen to be e.g., an hour; yet, to accommodate the rapidly changing renewable generation, the inverter output is adjusted (at the beginning of) every slot t, taken to be e.g., a minute. We assume that the values of $\mathbf{p}^g(\tau, t)$, $\mathbf{p}^c(\tau, t)$, and $\mathbf{q}^c(\tau, t)$ remain constant within per time slot t, although they may change between slots.

Let's consider a case where the distribution grid incorporates a total of N_a shunt capacitors. These capacitors are associated with specific bus indices, which we gather in the set N_a. To simplify our analysis, we establish a straightforward one-to-one correspondence between these capacitors and the elements of the set $\mathcal{K} := \{1, \ldots, N_a\}$ through a simple renumbering process. Furthermore, we assume that each bus in the grid is equipped with either a shunt capacitor or a smart inverter, but not both at the same time. After removing the entries in N_a from the original set N, the remaining buses, which we collect in the set N_r, are assumed to be equipped with inverters. It's important to emphasize that this assumption doesn't introduce any loss of generality in our analysis. To illustrate, we can easily enforce upper and lower bounds of zero on the reactive power output at buses that lack inverters, effectively accounting for their absence in our formulation.

Given that capacitor configuration operates on a relatively slow timescale, occurring every τ, we can express the reactive compensation, denoted as $q_i^g(\tau, t)$, provided by capacitor $k_i \in K$ (specifically, the capacitor at bus i), as follows

$$q_i^g(\tau, t) = \hat{y}_{k_i}(\tau) q_{a,k_i}^g, \forall i \in N_a, \tau, t, \tag{7.1}$$

where $\hat{y}_{k_i}(\tau) \in \{0, 1\}$ represents the on–off status of capacitor k_i for the entire interval τ. If $\hat{y}_{k_i}(\tau) = 1$, it indicates that a fixed amount of reactive power, denoted as q_{a,k_i}^g, is injected into the grid during this interval; otherwise, it is set to 0. To simplify the notation, we gather the on–off decisions for all capacitor units during interval τ into a column vector, denoted as $\hat{\mathbf{y}}(\tau)$.

On the other hand, the adjustment of reactive power $q_{r,i}^g(\tau, t)$ by inverter i occurs rapidly (every t), and it is subject to the constraint $|q_{r,i}^g(\tau, t)| \leq \sqrt{(\overline{s}_i)^2 - (p_i^g(\tau, t))^2}$, where $\overline{s}_i$ represents the power capability of inverter i. Conventionally, inverter i is designed with $\overline{s}_i = \overline{p}_i^g$, where $\overline{p}_i^g$ signifies the active power capacity of the renewable generation unit installed at bus i. However, when the maximum output is reached, i.e., $p_i^g(\tau, t) = \overline{p}_i^g$, it becomes impossible to provide any reactive power. To address this, oversized inverters' nameplate capacity has been advocated such that $\overline{s}_i > \overline{p}_i^g$ (Kekatos et al. 2014). Choosing $\overline{s}_i = 1.08\overline{p}_i^g$ and limiting $q_{r,i}^g(\tau, t)$ to $\sqrt{(\overline{s}_i)^2 - (\overline{p}_i^g)^2}$ instead of $\sqrt{(\overline{s}_i)^2 - (p_i^g(\tau, t))^2}$, the reactive power compensation provided by inverter i is $|q_{r,i}^g(\tau, t)| \leq 0.4\overline{p}_i^g$, regardless of the instantaneous PV output $p_i^g(\tau, t)$ (Kekatos et al. 2014). As such, $q_{r,i}^g(\tau, t)$ generated by inverter i is constrained as

$$\left|q_{r,i}^g(\tau, t)\right| \leq \overline{q}_r^g := \sqrt{(\overline{s}_i)^2 - (\overline{p}_i^g)^2}, \ \forall i \in \mathcal{N}_r, \tau. \tag{7.2}$$

7.2 Two-Timescale Voltage Regulation Formulation

Given the two-timescale load consumption and generation, which we model as Markovian processes (Carta et al. 2009), the objective of voltage regulation is to determine the optimal reactive power support for per time slot. This support involves configuring capacitors in every interval and adjusting inverter outputs in every slot. The primary goal is to minimize the long-term average voltage deviation. Voltage magnitudes $\mathbf{v}(\tau, t)$ are solely dependent on the control variables $\mathbf{q}^g(\tau, t)$ and are expressed as implicit functions of $\mathbf{q}^g(\tau, t)$, resulting in $\mathbf{v}_{\tau,t}(\mathbf{q}^g(\tau, t))$. The specific functional forms for these voltage magnitudes, based on the assumed grid models, will be provided in Sect. 7.3. The novel two-timescale voltage control scheme involves solving the following stochastic optimization problem

$$\min_{\substack{\mathbf{q}_r^g(\tau, t) \\ \mathbf{y}(\tau) \in \{0,1\}^{N_a}}} \quad \mathbb{E}\left[\sum_{\tau=1}^{\infty} \sum_{t=1}^{N_T} \gamma^T \left\|\mathbf{v}_{\tau,t}(\mathbf{q}^g(\tau, t)) - v_0 \mathbf{1}\right\|^2\right], \tag{7.3a}$$

$$\text{subject to} \quad q_i^g(\tau, t) = \hat{y}_{k_i}(\tau) q_{a,k_i}^g, \quad \forall i \in \mathcal{N}_a, \tau, t, \tag{7.3b}$$

$$q_i^g(\tau, t) = q_{r,i}^g(\tau, t), \quad \forall i \in \mathcal{N}_r, \tau, t, \tag{7.3c}$$

$$\left|q_{r,i}^g(\tau, t)\right| \leq \overline{q}_i^g, \quad \forall i \in \mathcal{N}_r, \tau, t, \tag{7.3d}$$

for some discount factor $\gamma \in (0, 1)$, where the expectation is taken over the joint distribution of $(\mathbf{p}^c(\tau, t), \mathbf{q}^c(\tau, t), \mathbf{p}^g(\tau, t))$ across all intervals and slots.

The optimization problem (7.3a)–(7.3d) presents a challenge due to its involvement of infinitely many variables $\left\{\mathbf{q}_r^g(\tau,t)\right\}$ and $\{\hat{\mathbf{y}}(\tau)\}$, which exhibit interconnections across time through the cost function and constraint (7.3b). Additionally, the presence of discrete variables $\{\hat{y}(\tau)\} \in \{0,1\}^{N_a}$ renders problem (7.3a)–(7.3d) nonconvex and generally categorized as NP-hard. Furthermore, it represents a multi-stage optimization, with decisions made at different stages, all while accommodating the variability in power during real-time operation. In essence, achieving an exact solution for (7.3a)–(7.3d) is a formidable task.

Our objective is to develop algorithms that iteratively observe predictions $\{(\mathbf{p}^c(\tau,t), \mathbf{q}^c(\tau,t), \mathbf{p}^g(\tau,t))\}$ and efficiently address problem (7.3a–7.3d). We operate under the assumption that, while we lack distributional knowledge of these stochastic processes, their realizations can be made available in real time through methods such as accurate forecasting techniques (Zhang et al. 2019b). In this context, we leverage the principles governing the electric power system, combined with data, to solve (7.3a–7.3d) in real-time. To elucidate, on the slow timescale, which occurs at the conclusion of each interval $\tau - 1$, the optimal decisions regarding the activation of capacitors, denoted as $\boldsymbol{y}(\tau)$, are determined using a DRL algorithm This algorithm learns from the predictions obtained during the preceding interval $\tau - 1$. Conversely, on the fast timescale, precisely at the commencement of per time slot t within interval τ, our two-stage control scheme calculates the optimal settings for inverters. This is achieved by minimizing instantaneous bus voltage deviations while adhering to physical constraints. These computations are based on the current on–off status of capacitor units $\hat{\mathbf{y}}(\tau)$, which are determined at the conclusion of interval $(\tau - 1)$. Further elaboration on these two timescales is provided in Sects. 7.3 and 7.4, respectively.

7.3 Fast-Timescale Optimization of Inverters

As previously mentioned, the specific expressions for $\mathbf{v}_{\tau,t}(\mathbf{q}^g(\tau,t))$ will be detailed in this section. These expressions are contingent upon the utilization of either the exact AC model or a linearized approximation of it. Employing convex relaxation to address nonconvexity, the AC model in question translates into a SOCP. Conversely, the linearized model results in a linearly constrained quadratic program. While the linearized model offers an approximation, it is computationally more economical compared to the exact AC model. The choice between these two models hinges on the available computational resources.

7.3.1 *Branch Flow Model*

Inherent to the radial structure of distribution grids is the presence of a unique parent bus π_i for every non-root bus i within the set $\mathcal{N}$. These buses are interconnected

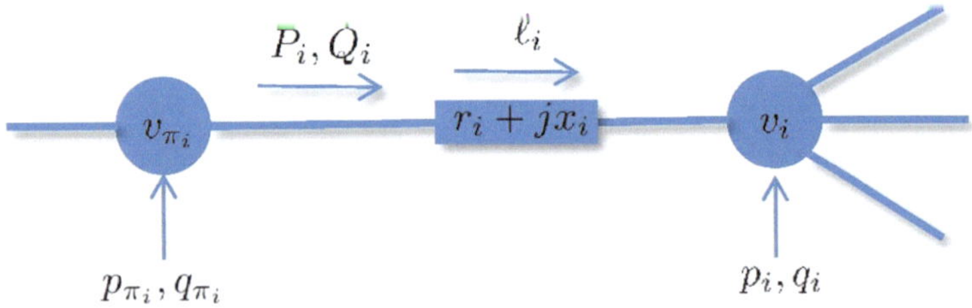

Fig. 7.2 Bus i is connected to its unique parent π_i via line i

via distribution lines denoted by $(\pi_i, i) \in \mathcal{L}$, each characterized by its impedance parameters $r_i + jx_i$. We represent the complex power flowing from the parent bus π_i to bus i at time slot t during interval τ as $P_i(\tau, t) + jQ_i(\tau, t)$. This configuration is illustrated in Fig. 7.2. In the subsequent sections, we will omit the interval index τ when it can be inferred from the context.

Additionally, we use ℓ_i to represent the squared magnitude of the current flowing through line $i \in \mathcal{L}$. The well-known branch flow model is defined by the following equations, applicable to all buses $i \in \mathcal{N}$ and for all t within each interval τ (Baran and Wu 1989a; Low 2014)

$$p_i(t) = \sum_{j \in \chi_i} P_j(t) - (P_i(t) - r_i \ell_i(t)), \tag{7.4a}$$

$$q_i(t) = \sum_{j \in \chi_i} Q_j(t) - (Q_i(t) - x_i \ell_i(t)), \tag{7.4b}$$

$$v_i(t) = v_{\pi_i}(t) - 2(r_i P_i(t) + x_i Q_i(t)) + (r_i^2 + x_i^2)\ell_i(t), \tag{7.4c}$$

$$\ell_i(t) = \frac{P_i^2(t) + Q_i^2(t)}{v_{\pi_i}(t)}, \tag{7.4d}$$

where we have ignored the dependence on τ for brevity, and χ_i denotes the set of all children buses for bus i.

The equations presented in (7.4d) are evidently quadratic in $P_i(t)$ and $Q_i(t)$, resulting in a nonconvex set. To tackle this complexity, we can explore the option of relaxing the equalities in (7.4d) into inequalities, a technique often referred to as hyperbolic relaxation (as described in Farivar et al. 2011)

$$P_i^2(t) + Q_i^2(t) \leq v_{\pi_i}(t)\ell_i(t), \ \forall i \in \mathcal{N}, t, \tag{7.5}$$

which can be equivalently rewritten as the following second-order cone constraints

$$\left\| \begin{matrix} 2P_i(t) \\ 2Q_i(t) \\ \ell_i(t) - v_{\pi_i}(t) \end{matrix} \right\| \leq v_{\pi_i}(t) + \ell_i(t), \ \forall i \in \mathcal{N}\ominus. \tag{7.6}$$

Equations (7.4a)–(7.4c) and (7.6) establish a convex feasible set. The approach of utilizing this relaxed set, as opposed to the nonconvex one, is commonly referred to as SOCP relaxation, as detailed in (Low 2014). Intriguingly, it has been demonstrated that under specific conditions, the SOCP relaxation is exact, meaning that the set of inequalities in (7.6) holds with equalities at the optimal solution (Farivar et al. 2011).

With the capacitor configuration $\hat{\mathbf{y}}(\tau)$ determined at the end of the previous interval $\tau - 1$, the voltage regulation on the fast timescale, utilizing the aforementioned relaxed grid model based on the exact AC model, can be described as follows

$$\min_{\mathbf{v}(t), \mathbf{q}_r^g(t), \mathbf{P}(t), \mathbf{Q}(t)} \quad \|\mathbf{v}(t) - v_0 \mathbf{1}\|^2, \tag{7.7a}$$

subject to, (7.4a)–(7.4d),

$$q_i^g(t) = \hat{y}_{k_i}(\tau) q_{a,k_i}^g, \quad \forall i \in \mathcal{N}_a, \tag{7.7b}$$

$$q_i^g(t) = q_{r,i}^g(t), \quad \forall i \in \mathcal{N}_r, \tag{7.7c}$$

$$\left| q_{r,i}^g(t) \right| \leq \overline{q}_i^g, \quad \forall i \in \mathcal{N}_r, \tag{7.7d}$$

which is readily a convex SOCP and can be efficiently solved by off-the-shelf convex programming toolboxes. The optimal setpoints for smart inverters in accordance with the exact AC model are determined by minimizing (7.7a)–(7.7d) with respect to $\mathbf{q}_r^g$.

Nonetheless, the computational demands of solving SOCPs can become substantial when applied to relatively large-scale distribution grids, such as those comprising several hundred buses. To strike a balance between modeling precision and computational efficiency, our subsequent approach for fast-timescale voltage control leans on an approximate grid model.

7.3.2 Linearized Power Flow Model

Given that line current magnitudes $\{\ell_i\}$ are considerably smaller in magnitude compared to line flows, it is reasonable to neglect the last term in (7.4a)–(7.4c), resulting in the following set of simplified linear equations for all i, t (Baran and Wu 1989b).

$$p_i(t) = \sum_{j \in \chi_i} P_j(t) - P_i(t), \tag{7.8a}$$

$$q_i(t) = \sum_{j \in \chi_i} Q_j(t) - Q_i(t), \tag{7.8b}$$

$$v_i(t) = v_{\pi_i}(t) - 2(r_i P_i(t) + x_i Q_i(t)), \tag{7.8c}$$

which is known as the linearized distribution flow model. In this manner, it is possible to express all squared voltage magnitudes $\mathbf{v}(t)$ as linear functions of $\mathbf{q}^g(t)$.

Utilizing the approximate model (7.8a)–(7.8c), the optimal setpoints for inverters can be determined by solving the following optimization problem for per time slot t within interval τ, assuming that $\hat{\mathbf{y}}(\tau)$ is available from the preceding interval on the slow timescale

$$\min_{\mathbf{v}(t), \mathbf{q}_r^g(t), \mathbf{P}(t), \mathbf{Q}(t)} \quad \|\mathbf{v}(t) - v_0 \mathbf{1}\|^2, \tag{7.9a}$$

subject to (7.8a)–(7.8c),

$$q_i^g(t) = \hat{y}_{k_i}(\tau) q_{a,k_i}^g, \quad \forall i \in \mathcal{N}_a, \tag{7.9b}$$

$$q_i^g(t) = q_{r,i}^g(t), \quad \forall i \in \mathcal{N}_r, \tag{7.9c}$$

$$\left| q_{r,i}^g(t) \right| \leq \overline{q}_i^g, \quad \forall i \in \mathcal{N}_r. \tag{7.9d}$$

Since all constraints are linear and the cost is quadratic, (7.9a)–(7.9d) forms a conventional convex quadratic program. Therefore, it can be efficiently solved using methods such as primal–dual algorithms or readily available convex programming solvers. Detailed implementation procedures are not provided here due to space constraints.

7.4 Slow-Timescale Capacitor Reconfiguration

In this section, we address the reconfiguration of shunt capacitors on the slow timescale. This involves determining whether these capacitors should be switched on or off for the upcoming interval. Previous methods for solving this integer-valued optimization problem were either heuristic or relied on SDP relaxation. These approaches do not guarantee optimality and often come with significant computational and storage complexities. Instead, we propose an innovative approach by leveraging recent advances in artificial intelligence. This allows us to develop data-driven solutions capable of learning, tracking, and adapting to unknown generation and consumption dynamics, providing near-optimal results.

7.4.1 A Data-Driven Solution

As evident from Eqs. (7.7b)–(7.9b), the decisions regarding capacitors, denoted as $\hat{\mathbf{y}}(\tau)$ and made at the end of interval $\tau - 1$ as part of the slow-timescale learning process, have a lasting impact on the setpoints of inverters throughout the entire interval τ as part of the fast-timescale optimization. Conversely, the regulation of voltages by inverters within the interval τ affects the commitment of capacitors for the subsequent interval. This intricate two-way relationship between capacitor configuration and the optimal setpoints of inverters provides the motivation for our RL formulation. RL deals with learning policy functions within an environment characterized by action-dependent, dynamically evolving states and associated costs. The primary objective of RL is to derive a policy function, based on the observed states, that generates actions aimed at minimizing the average cumulative cost (Sutton and Barto 2018).

Modeling load demand and renewable generation as Markovian processes, the optimal configuration of capacitors can be formulated as an Markovian decision processes, which can be efficiently solved through RL algorithms. An Markovian decision processes is defined as a 5-tuple $(\mathcal{S}, \mathcal{A}, \mathcal{P}, c, \gamma)$, where $\mathcal{S}$ is a set of states; $\mathcal{A}$ is a set of actions; $\mathcal{P}$ is a set of transition matrices; $c : \mathcal{S} \times \mathcal{P} \mapsto \mathbb{R}$ is a cost function such that, for $\mathbf{s} \in \mathcal{S}$ and $\mathbf{a} \in \mathcal{A}$, $c = (c(\mathbf{s}, \mathbf{a}))_{\mathbf{s} \in \mathcal{S}, \mathbf{a} \in \mathcal{A}}$ are the real-valued instantaneous costs after the system operator takes an action $\mathbf{a}$ at state $\mathbf{s}$; and $\gamma \in [0, 1)$ is the discount factor. These components are defined next before introducing our voltage regulation scheme.

Action space $\mathcal{A}$. Each action corresponds to a specific on–off commitment configuration for capacitors numbered from 1 to $\mathcal{N}_a$, giving rise to an action vector $\mathbf{a}(\tau) = \mathbf{y}(\tau)$ for each interval τ. This set of binary action vectors constitutes the action space $\mathcal{A}$, which displays exponential cardinality relative to the number of capacitors, implying that $|\mathcal{A}| = 2^{N_a}$.

State space $\mathcal{S}$. This comprises, for each interval τ, the average active power at all buses except for the substation, coupled with the current capacitor configurations. In other words, $\mathbf{s}(\tau) := [\overline{\mathbf{p}}^\top(\tau), \hat{\mathbf{y}}^\top(\tau)]^\top$, which encompasses both continuous and discrete variables. It is evident that $\mathcal{S}$ is a subset of $\mathbb{R}^N \times 2^{N_a}$.

The action is decided according to the configuration policy π that is a function of the most recent state $\mathbf{s}(\tau - 1)$, given as

$$\mathbf{a}(\tau) = \pi(\mathbf{s}(\tau - 1)). \tag{7.10}$$

Cost function c. The cost on the slow timescale is

$$c(\mathbf{s}(\tau - 1), \mathbf{a}(\tau)) = \sum_{t=1}^{N_T} \left\| \mathbf{v}_{\tau,t}(\mathbf{q}^g(\tau, t) - v_0 \mathbf{1}) \right\|^2. \tag{7.11}$$

Set of transition probability matrices $\mathcal{P}$. When the system is in a state $\mathbf{s} \in \mathcal{S}$ and an action $\mathbf{a}$ is taken, the system transitions probabilistically to a new state $\mathbf{s}' \in \mathcal{S}$.

We denote the transition probability matrix from state $\mathbf{s}$ to the subsequent state $\mathbf{s}'$ for a given action $\mathbf{a}$ as $P^{\mathbf{a}}_{\mathbf{ss}'}$. Clearly, the set of all such transition matrices is represented as $\mathcal{P} := \{P^{\mathbf{a}}_{\mathbf{ss}'} | \forall \mathbf{a} \in \mathcal{A}\}$.

Discount factor γ. The discount factor $\gamma \in [0, 1)$ holds a pivotal role in balancing present and future costs. A reduced value of γ imparts greater significance to the current cost in relation to future costs within the overarching optimization.

Given the current state and action, the so-called action-value function under the control policy π is defined as

$$Q_{\pi}(\mathbf{s}(\tau-1), \mathbf{a}(\tau)) := \mathbb{E}\left[\sum_{\tau'=\tau}^{\infty} \gamma^{\tau'-\tau} c(\mathbf{s}(\tau'-1), \mathbf{a}(\tau')) | \pi, \mathbf{s}(\tau-1), \mathbf{a}(\tau)\right], \tag{7.12}$$

where $\mathbb{E}$ is taken with respect to all sources of randomness.

To determine the optimal capacitor configuration policy π^*, with the objective of minimizing long-term average voltage deviation, we refer to the Bellman optimality equations, as elucidated in (Sutton and Barto 2018). Solving these equations yields the action-value function under the optimal policy π^*, which can be computed as follows

$$Q_{\pi^*}(\mathbf{s}, \mathbf{a}) := \mathbb{E}[c(\mathbf{s}, \mathbf{a})] + \gamma \sum_{\mathbf{s}' \in \mathcal{S}} P^{\mathbf{a}}_{\mathbf{ss}'} \min_{\mathbf{a} \in \mathcal{A}} Q_{\pi^*}(\mathbf{s}', \mathbf{a}'). \tag{7.13}$$

With $Q_{\pi^*}(\mathbf{s}, \mathbf{a})$ obtained, the optimal capacitor configuration policy can be found as

$$\pi^*(\mathbf{s}) = \arg\min_{\mathbf{a}} Q_{\pi^*}(\mathbf{s}, \mathbf{a}). \tag{7.14}$$

As shown in (7.13), it becomes evident that if we had access to all transition probabilities $\{P^{\mathbf{a}}_{\mathbf{ss}'}\}$, we could derive $Q_{\pi^*}(\mathbf{s}, \mathbf{a})$, and subsequently determine the optimal policy π^* using (7.14). However, obtaining these transition probabilities is unfeasible in practical distribution systems. There is a need to develop approaches that directly address π^* without making any assumptions about the knowledge of $\{P^{\mathbf{a}}_{\mathbf{ss}'}\}$.

One celebrated approach of this kind is Q-learning, which can learn π^* by approximating $Q_{\pi^*}(\mathbf{s}, \mathbf{a})$ 'on-the-fly' (Sutton and Barto 2018). Nevertheless, due to the high-dimensional and continuous nature of the state space $\mathcal{S}$, traditional Q-learning is not suitable for addressing the current problem. Consequently, this challenge has inspired the development of function approximation based Q-learning schemes designed to effectively handle continuous state domains.

7.4.2 A Deep Reinforcement Learning Approach

Deep Q-network (DQN) employs a NN function approximator for the $\mathcal{Q}$-function, with the choice of NN architecture depending on the specific application, such as a fully connected feed-forward NN or a convolutional NN (Mnih et al. 2015). This NN takes the state vector as input and produces $\mathcal{Q}$-values for all possible actions, one for each action. As demonstrated in prior research (Mnih et al. 2015), such a NN is capable of learning $\mathcal{Q}$-values for all state-action pairs effectively, even with a limited number of observations obtained through interactions with the environment. Consequently, it effectively addresses the challenge posed by the 'curse of dimensionality' (Mnih et al. 2015). Building upon this insight, we utilize a feed-forward NN to approximate the $\mathcal{Q}$-function within our framework. In particular, our DNNs comprises L fully connected hidden layers, each equipped with ReLU activation functions, as depicted in Fig. 7.3. At the input layer, each neuron receives one component of the state vector $\mathbf{s}(\tau - 1)$. This information propagates through the network's L ReLU layers, ultimately yielding an output vector $\mathbf{o}(\tau) \in \mathbb{R}^{2^{N_a}}$. The elements of this vector represent the predicted $\mathcal{Q}$-values for all feasible actions, namely, different configurations of the N_a capacitors. Since each output unit corresponds to a specific capacitor configuration, the output layer contains a total of 2^{N_a} neurons. For the sake of clarity, we gather all weight parameters of this DQN into a parameter vector $\boldsymbol{\theta}$, which characterizes the input–output mapping as $\mathbf{o}(\tau) = \mathcal{Q}_\pi(\mathbf{s}(\tau - 1), \mathbf{a}(\tau); \boldsymbol{\theta})$ (refer to Eq. (7.12)). At the conclusion of a given interval $\tau - 1$, after passing the state vector $\mathbf{s}(\tau - 1)$ through this DQN, we obtain the corresponding predicted $\mathcal{Q}$-values $\mathbf{o}(\tau)$ for all feasible actions at the output. Utilizing these predicted values, the system operator selects the action associated with the smallest predicted $\mathcal{Q}$-value, which will be implemented during the subsequent interval.

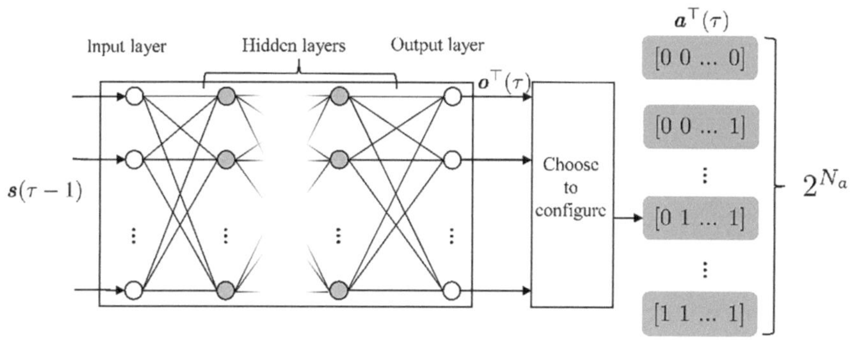

Fig. 7.3 Deep Q-network

In pursuit of our goal, the weights $\boldsymbol{\theta}$ are selected to ensure that the DQN produces accurate approximations of the true Q-values for any given state vector input. Toward this objective, the popular stochastic gradient descent (SGD) method is employed to update $\boldsymbol{\theta}$ 'on the fly' (Mnih et al. 2015). At the conclusion of each interval, precisely when: (i) The system operator has made the decision $\mathbf{a}(\tau)$. (ii) The grid has completed the transition from state $\mathbf{s}(\tau-1)$ to a new state $\mathbf{s}(\tau)$. (iii) The network has incurred and revealed the cost $c(\mathbf{s}(\tau-1), \mathbf{a}(\tau))$, we perform a SGD update to iteratively refine the parameter vector $\boldsymbol{\theta}_\tau$, resulting in $\boldsymbol{\theta}_{\tau+1}$. This process, often referred to as temporal-difference learning (Sutton and Barto 2018), is based on the principle that an approximate estimation of the optimal cost-to-go from interval τ is given by $c(\mathbf{s}(\tau-1), \mathbf{a}(\tau)) + \gamma \min_{\mathbf{a}' \in \mathcal{A}} Q_\pi(\mathbf{s}(\tau), \mathbf{a}'(\tau); \boldsymbol{\theta}_\tau)$, where $c(\mathbf{s}(\tau-1), \mathbf{a}(\tau))$ is the instantaneous cost observed, and $\min_{\mathbf{a}'} Q_\pi(\mathbf{s}(\tau), \mathbf{a}'(\tau); \boldsymbol{\theta}_\tau)$ represents the smallest possible predicted cost-to-go from state $\mathbf{s}(\tau)$, which can be computed through our DQN with weights $\boldsymbol{\theta}_\tau$, and is discounted by factor γ. In words, the target value $c(\mathbf{s}(\tau-1), \mathbf{a}(\tau)) + \gamma \min_{\mathbf{a}' \in \mathcal{A}} Q_\pi(\mathbf{s}(\tau), \mathbf{a}'(\tau); \boldsymbol{\theta}_\tau)$ is readily available at the end of interval $\tau-1$. To fine-tune the weights $\boldsymbol{\theta}$, we adopt the ℓ_2-norm error criterion, which involves minimizing the following loss function

$$\mathcal{L}(\boldsymbol{\theta}) := \left[c(\mathbf{s}(\tau-1), \mathbf{a}(\tau)) + \gamma \min_{\mathbf{a}' \in \mathcal{A}} Q_\pi(\mathbf{s}(\tau), \mathbf{a}'(\tau); \boldsymbol{\theta}_\tau) - Q_\pi(\mathbf{s}(\tau-1), \mathbf{a}(\tau); \boldsymbol{\theta})\right]^2 \tag{7.15}$$

for which the SGD update is given by

$$\boldsymbol{\theta}_{\tau+1} = \boldsymbol{\theta}_\tau - \beta_\tau \nabla \mathcal{L}(\boldsymbol{\theta})\big|_{\boldsymbol{\theta}_\tau}, \tag{7.16}$$

where $\beta_\tau > 0$ is a preselected learning rate, and $\nabla \mathcal{L}(\boldsymbol{\theta})$ denotes the (sub-) gradient.

However, in practice, straightforward application of the update rule (7.16) to DNNs often encounters challenges. The resulting DQN may exhibit instability and unreliable performance, as reported in previous studies (Wu et al. 2016). To overcome these issues, various enhancements have been introduced, and in this work, we employ two of these key modifications: the target network and experience replay, as proposed by (Mnih et al. 2015) To implement these improvements, we define an 'experience' at time $e(\tau') := (\mathbf{s}(\tau'-1), \mathbf{a}(\tau')), c(\mathbf{s}(\tau'-1), \mathbf{a}(\tau'), \mathbf{s}(\tau'))$. This experience is a tuple that includes the state, action, cost, and the resulting next state. In addition, we maintain an on-the-fly 'replay buffer' denoted as $\mathcal{R}(\tau)$, which stores the most recent $R > 0$ experiences encountered by the agent. For instance, at any interval $\tau \geq R$, the replay buffer is defined as $\mathcal{R}(\tau) := \{e(\tau-R+1), \ldots, e(\tau)\}$. Furthermore, as another effective remedy to stabilizing the DQN updates, we replicate the DQN to create a second DNNs, commonly referred to as the target network, whose weight parameters are concatenated in the vector $\boldsymbol{\theta}^{\text{Tar}}$. Notably, the target network is not actively trained, instead, its parameters $\boldsymbol{\theta}^{\text{Tar}}$ are periodically synchronized with the estimates from the primary DQN. This synchronization occurs, for example, every B training iterations of the primary DQN.

Consider now the temporal-difference loss for some randomly drawn experience $e(\tau')$ from $\mathcal{R}(\tau)$ at interval τ

$$\mathcal{L}^{\text{Tar}}(\boldsymbol{\theta}_\tau; e(\tau')) := \frac{1}{2}[c(\mathbf{s}(\tau'-1), \mathbf{a}(\tau')) + \gamma \min_{\mathbf{a}'} Q^{\text{Tar}}(\mathbf{s}(\tau), \mathbf{a}'; \boldsymbol{\theta}_{\tau'}^{\text{Tar}}) \\ - Q(\mathbf{s}(\tau'-1), \mathbf{a}(\tau'); \boldsymbol{\theta}_\tau)]^2. \tag{7.17}$$

Upon taking expectation with respect to all sources of randomness generating this experience, we arrive at

$$\mathcal{L}^{\text{Tar}}(\boldsymbol{\theta}_\tau; \mathcal{R}(\tau)) := \mathbb{E}_{e(\tau')}\mathcal{L}^{\text{Tar}}(\boldsymbol{\theta}_\tau; e(\tau')). \tag{7.18}$$

However, in practical applications, the precise values of these transition probabilities are typically unknown, making it challenging to evaluate and minimize $\mathcal{L}^{\text{Tar}}(\boldsymbol{\theta}_\tau; \mathcal{R}(\tau))$ directly. As an alternative, a commonly adopted approach is to approximate the expected loss with an empirical loss computed over a limited set of samples, which in this context are experiences. To implement this approximation, we randomly select a mini-batch of M_τ experiences from the replay buffer $\mathcal{R}(\tau)$, where the indices of these experiences are collected in the set $\mathcal{M}_\tau$, formally denoted as $\{e(\tau')\}_{\tau' \in \mathcal{M}_\tau} \sim U(\mathcal{R}(\tau))$. For each of these sampled experiences, we use the target network with parameters $\boldsymbol{\theta}_\tau^{\text{Tar}}$ to compute an output. The empirical loss is then calculated based on these outputs and is given by

$$\mathcal{L}^{\text{Tar}}(\boldsymbol{\theta}_\tau; \mathcal{M}_\tau) := \frac{1}{2M_\tau} \sum_{\tau' \in \mathcal{M}_\tau} [c(\mathbf{s}(\tau'-1), \mathbf{a}(\tau')) + \gamma \min_{\mathbf{a}'} Q^{\text{Tar}}(\mathbf{s}(\tau), \mathbf{a}'; \boldsymbol{\theta}_{\tau'}^{\text{Tar}}) \\ - Q(\mathbf{s}(\tau'-1), \mathbf{a}(\tau'); \boldsymbol{\theta}_\tau)]^2. \tag{7.19}$$

In a nutshell, the weight parameter vector $\boldsymbol{\theta}_\tau$ of the DQN is efficiently updated 'on-the-fly' using SGD over the empirical loss $\mathcal{L}^{\text{Tar}}(\boldsymbol{\theta}_\tau; \mathcal{M}_\tau)$, with iterates given by

$$\boldsymbol{\theta}_{\tau+1} = \boldsymbol{\theta}_\tau - \beta_\tau \nabla \mathcal{L}^{\text{Tar}}(\boldsymbol{\theta}_\tau; \mathcal{M}_\tau). \tag{7.20}$$

Incorporating target network and experience replay remedies for stable DRL, our proposed two-timescale voltage regulation scheme is summarized in Algorithm 1.

Algorithm 1 Two-timescale voltage regulation scheme

Initialize: $\boldsymbol{\theta}_0$ randomly; weight of the target network $\boldsymbol{\theta}_0^{\mathrm{Tar}} = \boldsymbol{\theta}_0$; replay buffer $\mathcal{R}$; and the initial state $\mathbf{s}(0)$
for $\tau = 1, 2, \ldots$ **do**
Take action $\mathbf{a}(\tau)$ through exploration–exploitation

$$\mathbf{a}(\tau) = \begin{cases} \text{random } \mathbf{a} \in \mathcal{A} \text{ w.p.} \varepsilon_\tau \ominus \\ \arg\min_{\mathbf{a}'} Q(\mathbf{s}(\tau-1), \mathbf{a}'; \boldsymbol{\theta}_\tau) \text{ w.p.} 1-\varepsilon_\tau \end{cases}$$

where $\varepsilon_\tau = \max\{1 - 0.1 \times \lfloor \tau/50 \rfloor, 0\}$
Evaluate $\mathbf{c}(\mathbf{s}(\tau-1), \mathbf{a}(\tau))$ using (7.11)
for $t = 1, 2, \ldots N_T$ **do**
Compute $\mathbf{q}^g(\tau, t)$ using (7.7a–(7.7d) or (7.9a)–(7.9d)
end for
Update $\mathbf{s}(\tau)$
Save $(\mathbf{s}(\tau-1), \mathbf{a}(\tau), c(\mathbf{s}(\tau-1), \mathbf{a}(\tau)), \mathbf{s}(\tau))$ into $\mathcal{R}(\tau)$
Randomly sample M_τ experiences from $\mathcal{R}(\tau)$
Form the mini-batch loss $\mathcal{L}^{\mathrm{Tar}}(\boldsymbol{\theta}_\tau; \mathcal{M}_\tau)$ using (7.19)
Update $\boldsymbol{\theta}_{\tau+1}$ using (7.20)
if mod $(\tau, B) = 0$ **then**
Update the target network $\boldsymbol{\theta}_\tau^{\mathrm{Tar}} = \boldsymbol{\theta}_\tau$
end if
end for

7.5 Numerical Tests

In this section, we present numerical tests conducted on a real-world 47-bus distribution feeder and the IEEE 123-bus benchmark system to demonstrate the effectiveness of our proposed DRL-based voltage control scheme (refer to Algorithm 1). As established in prior studies (e.g., Kekatos et al, 2014; Kekatos et al. 2015; Low 2014), the linearized distribution flow model provides an accurate approximation of the exact AC model. Consequently, we focus our numerical results on the linearized model in this section.

The initial experiment revolves around the Southern California Edison 47-bus distribution feeder, depicted in Fig. 7.4 and thoroughly detailed in (Farivar et al. 2011). This feeder is equipped with four shunt capacitors and five intelligent inverters. Notably, the voltage magnitude v_0 at the substation bus remains constant (set at 1 in all our tests) due to the deployment of a voltage transformer. The substation's capacitor has been excluded from our control strategy, resulting in a voltage regulation scheme that centers on three shunt capacitors and five intelligent inverters, all integrated with sizable PV plants. The remaining three capacitors are strategically positioned on buses 3, 37, and 47, boasting capacities of 120, 180, and 180 kVar, respectively. Simultaneously, the five substantial PV plants are situated on buses 2, 16, 18, 21,

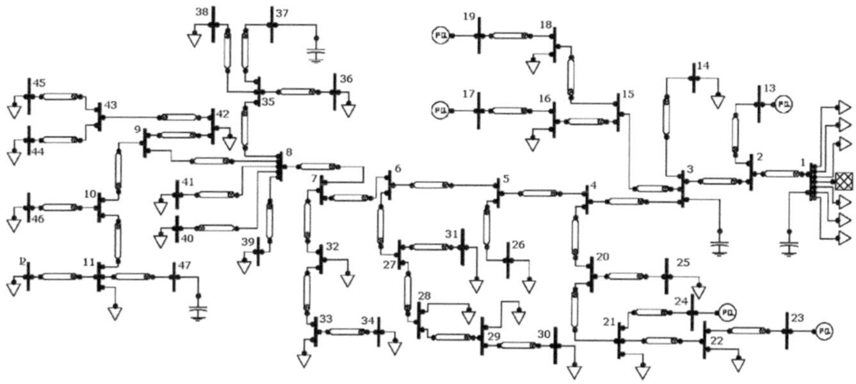

Fig. 7.4 Schematic diagram of the 47-bus industrial distribution feeder. Bus 1 is the substation, and the 6 loads connected to it model other feeders on this substation

and 22, with capacities of 300, 80, 300, 400, and 200 kW, respectively. To assess the performance of our approach in a realistic context, we leveraged actual consumption and solar generation data from the Smart * project, specifically collected on August 24, 2011, as documented in (Barker et al. 2012). These data underwent preprocessing, following the outlined procedure in our previous work (Kekatos et al. 2014).

In our experiments, we configured per time slot t to represent one minute, with each interval τ corresponding to five minutes, aligning with the temporal granularity of the real data. Throughout the simulations, we assumed a power factor of 0.8 for all loads. The DQN employed in our experiments featured three fully connected layers, comprising 44 units in the first hidden layer and 12 units in the second hidden layer. Despite its simplicity, this architecture demonstrated effectiveness for the given task. We applied ReLU activation functions ($\sigma(x) = \max(x, 0)$) in the hidden layers and logistic sigmoid functions $s(x) = 1/(1 + e^{-x})$ at the output layer. To evaluate the effectiveness of our proposed scheme, we conducted simulations with three baseline capacitor configuration policies: a fixed capacitor configuration (FixCap), a random capacitor configuration (RandCap), and a computationally impractical 'real-time' policy. The FixCap policy maintains a constant capacitor configuration throughout the simulation, while the RandCap policy randomly selects capacitor configurations at each slow time interval. Both of these policies calculate inverter setpoints by solving (7.9a)–(7.9d) for per time slot t. In contrast, the impractical Real-time policy optimizes both inverters and capacitors on a single timescale, specifically at every time slot. To perform this optimization, we initially relax the binary constraints $y_{k_i}(t) \in \{0, 1\}$ to continuous ones $y_{k_i}(t) \in [0, 1]$. We then solve the resulting convex program using a standard optimization routine (Grant and Boyd 2014). Subsequently, a standard rounding step is applied to recover binary solutions for capacitor configurations (Baran and Wu 1989c).

In the first experiment, the DRL based capacitor configuration (DRLCap) voltage control approach was examined. The replay buffer size was set to $R = 10$, the discount factor $\gamma = 0.99$, the mini-batch size $M_\tau = 10$, and the exploration–exploitation parameter $\varepsilon_\tau = \max\{1 - 0.1 \times \lfloor \tau/50 \rfloor, 0\}$ During training, the target network was updated every $B = 5$ iterations. The time-averaged instantaneous costs $\frac{1}{\tau}\sum_{i=1}^{\tau} c(\mathbf{s}(i-1), \mathbf{a}(i))$ incurred by the four schemes over the first $1 \le \tau \le 2,000$ intervals are plotted in Fig. 7.5. Evidently, the proposed scheme attains a lower cost than FixCap, RandCap, and Real-time after a short period of learning and interacting with the environment. Despite the real-time scheme's attempt to optimize both capacitor configurations and inverter setpoints on a per-slot basis, its suboptimal performance in this context can be attributed to the significant disparity between the convexified problem and its original nonconvex counterpart. To illustrate this, we examine the voltage magnitude profiles for all buses regulated by the four schemes, sampled at every 100 slots, as shown in Fig. 7.6. Once again, we observe that after a relatively brief training period (approximately 4500 slots) involving interaction with the environment, our DRLCap voltage control scheme rapidly adapts to a stable and nearly optimal policy. Further insight into its effectiveness is provided by examining the voltage magnitude profiles at buses 10 and 33 from slot 9900 to 10000, as depicted in Fig. 7.7. Additionally, Fig. 7.8 presents the voltage magnitude profiles at all buses at slot 9900. These profiles collectively highlight the remarkable ability of our DRLCap scheme to mitigate voltage fluctuations resulting from both substantial solar generation and heavy load demand.

To deal with distribution systems having a moderately large number of capacitors, we further advocate a hyper deep Q-network implementation, that endows our DRL-based scheme with scalability. The idea here is to first split the total number 2^{N_a} of Q-value predictions $\mathbf{o}(\tau) \in \mathbb{R}^{2^{N_a}}$ at the output layer into K smaller groups, each of which is of the same size $2^{N_a}/K$ and is to be predicted by a small-size DQN. This evidently yields the representation $\mathbf{o}(\tau) := [\mathbf{o}_1^\top(\tau), \ldots, \mathbf{o}_K^\top(\tau)]^\top$, where $\mathbf{o}_k(\tau) \in \mathbb{R}^{2^{N_a}/K}$ for

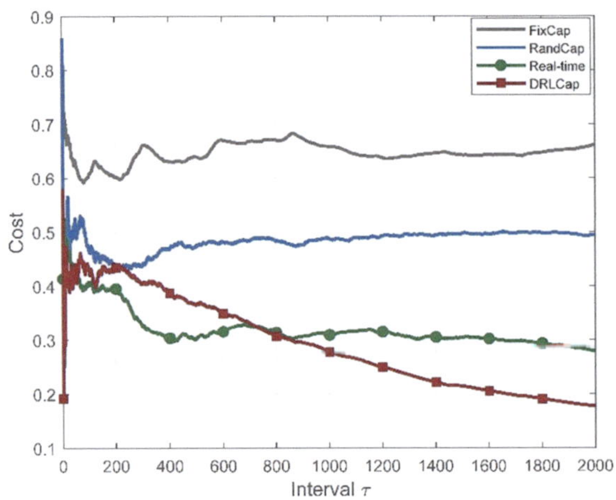

Fig. 7.5 Time-averaged instantaneous costs incurred by the four voltage control schemes

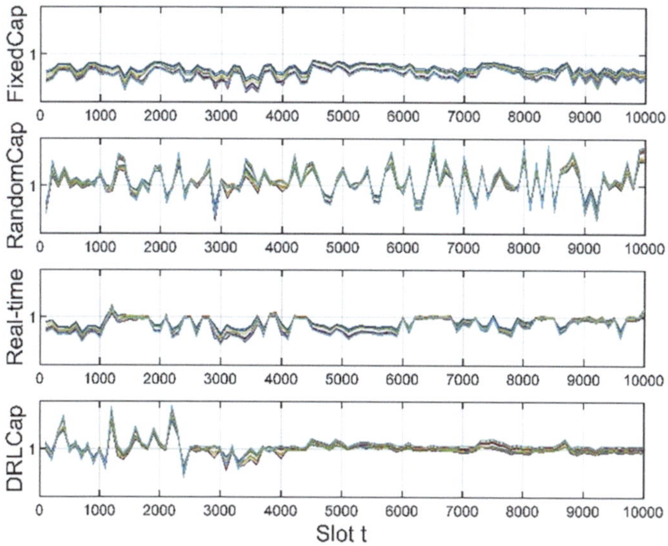

Fig. 7.6 Voltage magnitude profiles obtained by the four voltage control schemes over the simulation period of 10,000 slots

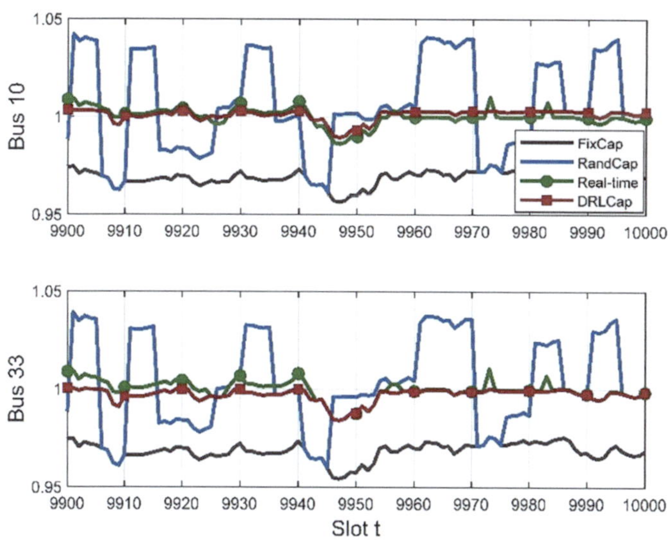

Fig. 7.7 Voltage magnitude profiles obtained by the four voltage control schemes at buses 10 and 33 from slot 9900 to 10,000

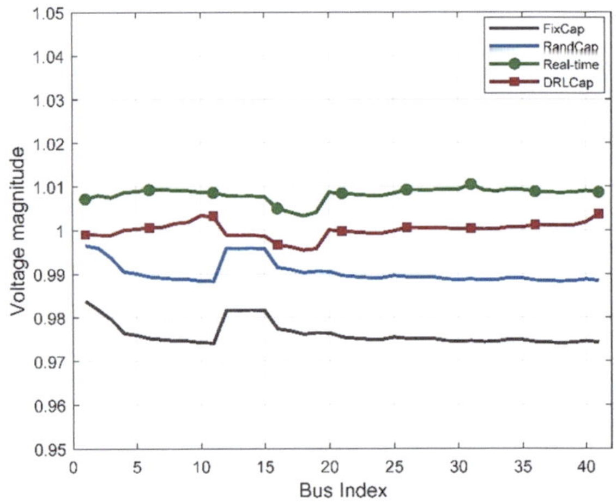

Fig. 7.8 Voltage magnitude profiles at all buses at slot 9900 obtained by the four voltage control schemes

$k = 1, \ldots, K$. By running K DQNs in parallel along with their corresponding target networks, each DQN-k generates predicted Q-values $\mathbf{o}_k(\tau)$ for the subset of actions corresponding to k th group. Note that all DQNs are fed with the same state vector $\mathbf{s}(\tau - 1)$; see also Fig. 7.9 for an illustration.

To assess the scalability and performance of the hyper Q-network implementation, tests were conducted using the IEEE 123-bus test feeder, which incorporates 9 shunt capacitors. Similar to previous experiments, the capacitor at bus 1 was excluded from

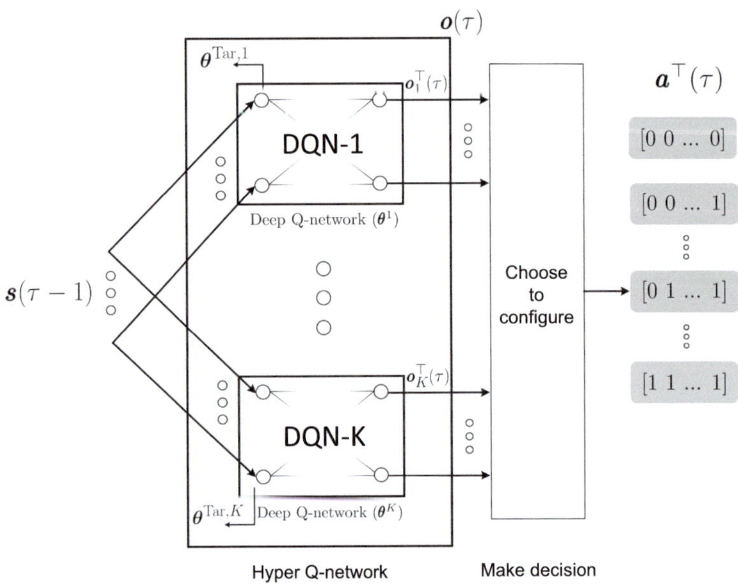

Fig. 7.9 Hyper deep Q-network for capacitor configuration

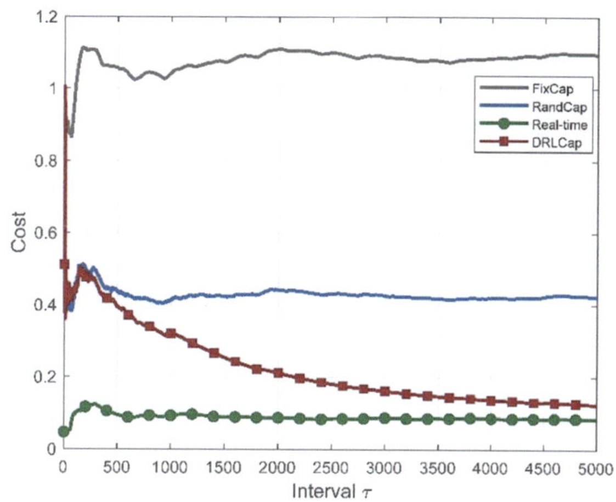

Fig. 7.10 Time-averaged instantaneous costs incurred by the four approaches on the IEEE 123-bus feeder

the control, resulting in a total of 256 possible actions (capacitor configurations). PV units were strategically positioned on buses 47, 49, 63, 73, 104, 108, and 113, each with specific capacities of 100, 16, 70, 20, 20, 30, and 10 kVA, respectively. Shunt capacitors were installed on buses3, 20, 44, 93, 96, 98, 100, and 114, with respective capacities of 50, 80, 100, 100, 100, 100, 100, and 60 kVar in this experiment. A total of 64 equally sized DQNs were employed to construct the hyper Q-network. Each of these DQNs was implemented as a fully connected 3-layer FFNNs, featuring ReLU activation functions in the hidden layers and sigmoid functions at the output.

In our IEEE 123-bus test experiments, we configured the replay buffer with a size of $R = 50$, utilized a batch size of $M_\tau = 8$, and updated the target network every $B = 10$ iterations. We conducted simulations over a period of 5000 intervals and analyzed the results as follows. First, in Fig. 7.10, we present the time-averaged instantaneous costs obtained throughout the simulation. Specifically, we provide voltage magnitude profiles for all buses at slot 24,900 in Fig. 7.11. Evidently, our hyper deep Q-network based DRL scheme significantly reduces voltage fluctuations after a learning period of approximately 7000 slots, effectively addressing the challenge posed by the high dimensionality of the control (action) space. As observed in both Figs. 7.10 and 7.11, the time-averaged immediate cost and the voltage profiles achieved by DRLCap converge to those of the impractical 'real-time' scheme, which jointly optimizes inverter setpoints and capacitor configurations per slot. This convergence demonstrates the effectiveness of our proposed DRL-based voltage control approach in achieving near-optimal performance.

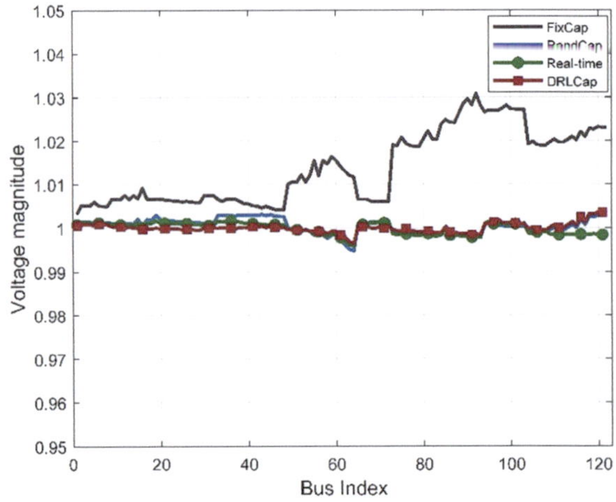

Fig. 7.11 Voltage magnitude profiles at all buses on slot 24,900 obtained by four approaches on the IEEE 123-bus feeder

Chapter 8
Learning Control Polices via Policy Gradients

The reliability and operational efficiency of contemporary distribution systems face formidable challenges due to the substantial integration of unpredictable renewable energy resources, the extensive adoption of electric vehicles, and the incorporation of 'human-in-the-loop' demand response initiatives. Consequently, reverse power flow and voltage magnitude fluctuations have become prevalent phenomena within nowadays residential grids (Carvalho et al. 2008). In addressing the challenges posed by renewable energy uncertainties and the unreliable communication links that introduce delays and potential communication failures, researchers have developed stochastic, online, decentralized, and localized smart inverter control schemes (Kekatos et al. 2014; Lin et al. 2018; Zhu and Liu 2015). However, it's worth noting that centralized solvers tend to grapple with significant computational complexity, while decentralized and localized algorithms exhibit relatively slower convergence rates (Lin et al. 2018; Sun et al. 2019; Lin et al. 2018).

To overcome these challenges, recent advancements have harnessed machine learning techniques to facilitate rapid networked control and monitoring (Yang et al. 2019; Zhang et al. 2019b; Zamzam and Sidiropoulos 2020; Cao et al. 2020). One notable approach involved the development of a support vector machine-based method in (Jalali et al. 2019) to approximate an almost optimal inverter control strategy. Similar approaches harnessing DNNs that 'learn-to-optimize' have emerged in various domains, including resource allocation (Lee et al. 2018), optimal power flow (Zamzam and Baker 2020; Pan et al. 2020; Hu et al. 2020), and PSSE (Zhang et al. 2019b; Zamzam and Sidiropoulos 2020). However, it's worth noting that training conventional supervised learning models for reactive power control often demands a substantial amount of labeled training data, which can be challenging to acquire in real-world physical systems. Contrastingly, RL approaches necessitate prior knowledge for designing reward functions and frequently demonstrate slower convergence rates.

This chapter adopts an approach rooted in statistical learning, harnessing the power of DNNs to parameterize the input–output relationships linking diverse

G. Wang et al., *Data-driven Optimization and Control for Autonomous Energy Systems*,
https://doi.org/10.1007/978-981-95-1782-4_8

power grid states with their corresponding optimal reactive power control strategies. Through the utilization of the policy gradient method, the NNs' unknown weights undergo iterative updates aimed at minimizing the cumulative power loss across a multitude of historical and simulated training pairs. During the inference phase, simply inputting real-time state vectors into the learned NNs allows for the expeditious derivation of 'optimal' reactive power control decisions within the computational confines of just a few matrix–vector multiplications. The salient advantages of this pioneering deep policy gradient method encompass its heightened computational efficiency and robustness in the face of stochastic input perturbations. These practical advantages have been empirically confirmed through numerical experiments conducted on a 47-bus distribution network using actual solar energy and consumption data.

8.1 System Modeling and Problem Formulation

Consider a radial power distribution network modeled by a tree graph $\mathcal{G} := (\mathcal{N}_0, \mathcal{E})$, where $\mathcal{N}_0 := \{0\} \cup \mathcal{N}$ denotes the set of buses, and $\mathcal{E}$ the set of edges. The tree is rooted at the substation bus indexed by $n = 0$, and all branch buses are collected in $\mathcal{N} := \{1, 2, \ldots, N\}$. For each bus $i \in \mathcal{N}$, let v_n denote its squared voltage magnitude, and $p_n + jq_n$ denote its complex power injection, where $p_n := p_n^g - p_n^c$ and $q_n := q_n^g - q_n^c$ with superscript ^{g} (^{c}) specifying generation (consumption).

The radial topology of the distribution grid offers a convenient structure, ensuring that each non-root bus $n \in \mathcal{N}$ possesses a unique parent bus, denoted as π_n. These buses are interconnected through the distribution lines $(\pi_n, n) \in \mathcal{E}$, with each line characterized by its impedance, expressed as $r_n + jx_n$. We use $P_n + jQ_n$ to represent the complex power flow from the parent bus π_n to bus n, as observed at the 'front' end of the line. Additionally, ℓ_n signifies the squared magnitude of the current flowing through line n, belonging to the set $\mathcal{E}$. For ease of reference, we organize all nodal and line parameters into column vectors denoted as $\mathbf{v}$, $\mathbf{p}$, $\mathbf{q}$, $\mathbf{p}^g$, $\mathbf{q}^g$, $\mathbf{p}^c$, $\mathbf{q}^c$, and 1. Refer to Fig. 8.1 for an illustrative diagram of this configuration.

The radial grid can be succinctly characterized using the well-established branch flow model 1 (Baran and Wu 1989a). This model imposes a set of equations that govern the behavior of all buses n within the network, expressed as follows

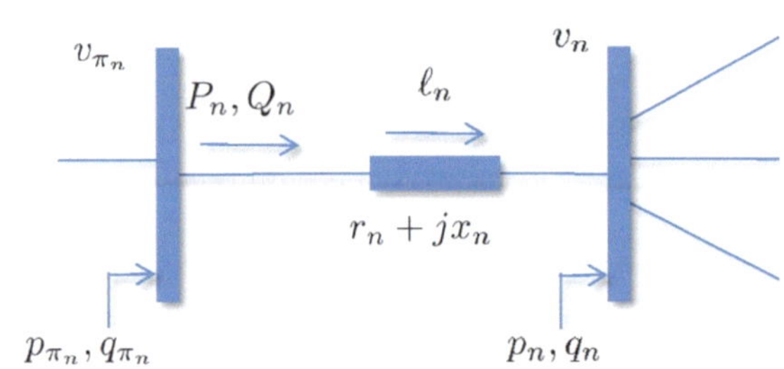

Fig. 8.1 Bus n is connected to its unique parent π_n via line n

$$P_n = \sum_{j \in \chi_n} P_j - p_n + r_n \ell_n, \tag{8.1a}$$

$$Q_n = \sum_{j \in \chi_n} Q_j - q_n + x_n \ell_n, \tag{8.1b}$$

$$v_n = v_{\pi_n} - 2(r_n P_n + x_n Q_n) + (r_n^2 + x_n^2)\ell_n, \tag{8.1c}$$

$$\ell_n = \frac{P_n^2 + Q_n^2}{v_{\pi_n}}, \tag{8.1d}$$

where the set $\chi \subseteq \mathcal{N}$ collects all children buses for bus n.

Traditionally, when considering a smart inverter located at bus n with a nominal power capacity denoted as $\bar{s}_n$, and a solar panel installed at the same bus, featuring a nameplate active power capacity denoted as $\bar{p}_n^g$, it is customary to set $\bar{s}_n = \bar{p}_n^g$. Moreover, there exist constraints on the reactive power q_n^g generated by the inverter, defined as follows $|q_n^g| \leq \sqrt{(\bar{s}_n)^2 - (p_n^g)^2}$, $\forall n$, where p_n^g is the smart inverter output. However, to capture the special case that no reactive power can be provided when the maximum inverter output is reached (i.e., $p_n^g = \bar{p}_n^g$), oversized inverters' nameplate capacity (i.e., $\bar{s}_n > \bar{p}_n^g$) is used in practice Turitsyn et al. (2011).

For example, when considering the reactive power compensation provided by inverter n, it is possible to set the constraint as $|q_n^g| \leq 0.4\bar{p}_n^g$. This can be achieved by choosing $\bar{s}_n = 1.08\bar{p}_n^g$ and limiting q_n^g to $\sqrt{(\bar{s}_n)^2 - (\bar{p}_n^g)^2}$ instead of $\sqrt{(\bar{s}_n)^2 - (p_n^g)^2}$, irrespective of the instantaneous PV output p_n^g (Kekatos et al. 2014). Under this policy, the region for reactive power injection remains time-invariant and takes the form of a convex set

$$\underline{\mathbf{q}}^g \leq \mathbf{q}^g \leq \overline{\mathbf{q}}^g, \tag{8.2}$$

where $\mathbf{q}^g \in \mathcal{R}^M$, and M denotes the number of inverters in the grid. Moreover, the voltage magnitude at every bus $n \in \mathcal{N}$ should be maintained within a prespecified range, i.e., $v_n \in [\underline{v}_n, \bar{v}_n]$. In practice this range is chosen to be $\pm 5\%$ of its nominal value. For future use, rewrite voltage regulation constraints at all buses $n \in \mathcal{N}$ in a compact way as

$$\mathbf{v} \in \mathcal{V} := \{\underline{\mathbf{v}} \leq \mathbf{v} \leq \overline{\mathbf{v}}\}. \tag{8.3}$$

In distribution grids, the power characteristics of each bus type are as follows: When bus n only has a capacitor, it satisfies $p_n^g = p_n^c = q_n^c = 0$ and $q_n^g > 0$. For a purely load bus n, it holds that $p_n^g = q_n^g = 0$, $p_n^c \geq 0$, and $q_n^c \geq 0$. In this case, the bus consumes active power denoted by p_n^c and reactive power denoted by q_n^c. A DG bus n not only consumes power, specified by p_n^c and q_n^c, but also generates active power $p_n^g \geq 0$. Additionally, it can provide either negative or positive reactive power q_n^g. Furthermore, predictions for active power consumption and solar generation, represented by the vectors $(\mathbf{p}^c, \mathbf{q}^c, \mathbf{p}^g)$, can be obtained through hourly and real-time market data sources (see, for instance, Kekatos et al. 2014), or by taking advantage of load demand and solar generation prediction algorithms (Zhang et al. 2019b).

8.2 Problem Formulation

In the envisioned distribution network operational case, the regulation of active power $\mathbf{p}$ takes place at a relatively coarse timescale. Depending on factors such as the variability of active power and the availability of cyber resources (including sensing, communication, and computation delays), the compensation of reactive power occurs over discrete time intervals, denoted by $t = 0, 1, \ldots$. These intervals can correspond to real-time market periods, such as every 5 min, or even shorter periods, such as 30 s. Let $(\mathbf{p}_t, \mathbf{q}_t)$ represent the active and reactive power injections at all non-root buses during control period t. The total power loss across all distribution lines can be succinctly expressed as $\sum_{n=1}^{N} r_{n,t}\ell_{n,t}$. Given the initial load consumptions $(\mathbf{p}_t^c, \mathbf{q}_t^c)$ and the available generation $\mathbf{p}_t^g$ at the start of per time interval t, the objective of reactive power control is to determine feasible reactive power injections $\mathbf{p}_t^{g,*}$ for smart inverters. This control strategy aims to minimize the power losses across all distribution lines while ensuring that all bus voltage magnitudes remain within a specified range. To formalize this, the reactive power control problem is defined as follows

$$\mathbf{q}_t^{g,*} := \arg \min_{\underline{\mathbf{q}}^g \le \mathbf{q}^g \le \overline{\mathbf{q}}^g} f(\mathbf{p}_t, \mathbf{q}^g - \mathbf{q}_t^c), \tag{8.4}$$

where $f(\mathbf{p}_t, \mathbf{q}^g - \mathbf{q}_t^c)$ admits the following form

$$f(\mathbf{p}_t, \mathbf{q}^g - \mathbf{q}_t^c) = \min_{\substack{\mathbf{P}_t, \mathbf{Q}_t \\ \mathbf{l}_t, \mathbf{v}_t}} \sum_{n=1}^{L} r_{n,t}\ell_{n,t}, \tag{8.5a}$$

$$\text{subject to}\, P_{n,t} = \sum_{j \in \mathcal{C}_{n,t}} P_{j,t} - p_{n,t} + r_{n,t}\ell_{n,t},\ n \in \mathcal{N}, \tag{8.5b}$$

$$Q_{n,t} = \sum_{j \in \mathcal{C}_n} Q_{j,t} - q_{n,t} + x_{n,t}\ell_{n,t},\ n \in \mathcal{N}, \tag{8.5c}$$

$$v_{n,t} = v_{\pi_{n,t}} + (r_{n,t}^2 + x_{n,t}^2)\ell_{n,t} - 2(r_{n,t}P_{n,t} + x_{n,t}Q_{n,t}),\ n \in \mathcal{N}, \tag{8.5d}$$

$$\ell_{n,t} = \frac{P_{n,t}^2 + Q_{n,t}^2}{v_{\pi_{n,t}}},\ \mathrm{n} \in \mathcal{E}, \tag{8.5e}$$

$$\mathbf{v} \in \mathcal{V}. \tag{8.5f}$$

Clearly, constraints (8.5b–8.5d) and (8.5f) are linear concerning the system variables $(\mathbf{p}_t, \mathbf{q}_t, \mathbf{P_t}, \mathbf{Q_t}, \mathbf{l_t}, v_\mathbf{t})$. However, the constraints in (8.5e) take the form of quadratic equalities, defining a non-convex feasible set that inherently makes the optimization problem non-convex and, in general, NP-hard (Low 2014).

To address this problem, the equalities in (8.5e) have been replaced with convex inequalities using hyperbolic constraints (Low 2014).

$$P_{n,t}^2 + Q_{n,t}^2 \leq v_{\pi_{n,t}} \ell_{n,t}. \tag{8.6}$$

Substituting (8.6) into (8.5b) and (8.5c) yields

$$f(\mathbf{p}_t, \mathbf{q}_t) = \min_{\substack{\mathbf{P}_t, \mathbf{Q}_t \\ \mathbf{l}_t, \mathbf{v}_t}} \sum_{n=1}^{L} r_{n,t} \ell_{n,t}, \tag{8.7a}$$

subject to (8.5b–8.5d) and (8.5f),

$$\ell_{n,t} \geq \frac{P_{n,t}^2 + Q_{n,t}^2}{v_{\pi_{n,t}}}, \quad \mathrm{n} \in \mathcal{E}, \tag{8.7c}$$

where (8.7c) can also be equivalently expressed as a second-order cone

$$\left\| \begin{matrix} 2P_{n,t} \\ 2Q_{n,t} \\ \ell_{n,t} - v_{\pi_{n,t}} \end{matrix} \right\| \leq v_{\pi_{n,t}} + \ell_{n,t}. \tag{8.8}$$

Constraints (8.7b) and (8.7c) now define a convex feasible set, making the problem in (8.7a–8.7c) amenable to standard convex programming techniques. Intriguingly, it has been demonstrated that under specific conditions, the optimal solution of (8.7a–8.7c) satisfies the equalities in (8.8); for more details, refer to Low, 2014. In such cases, the optimal solution of the original problem (8.5a–8.5f) is also obtained.

It's noteworthy that problem (8.4) formally delineates the optimal reactive power control strategies for a wide array of networked control problems, encompassing voltage regulation, Volt/VAR control, and optimal power flow, among others (Low 2014), contingent upon the choice of suitable objective functions. If both active and reactive power injections $(\mathbf{p}_t, \mathbf{q}_t^c)$ were precisely known in advance and remained constant within period t, determining the optimal reactive power compensation $\mathbf{q}_t^{g,*}$ would involve solving (8.4). However, contemporary distribution systems rarely conform to such ideal conditions due, in part, to (i) time-varying active and reactive injections, and (ii) noise-introduced discrepancies arising from direct measurements, delayed estimations, or inaccurate forecasts. In light of these challenges, an alternative to the static reactive power control formulation in (8.4) is to minimize the average power loss over the power injections $(\mathbf{p}_t, \mathbf{q}_t^c)$, as given by

$$\mathbf{q}_t^{g,*} := \arg \min_{\underline{\mathbf{q}}^g \leq \mathbf{q}^g \leq \overline{\mathbf{q}}^g} \mathbb{E}\big[f(\mathbf{p}_t, \mathbf{q}^g - \mathbf{q}_t^c)\big]. \tag{8.9}$$

To facilitate notation, we can introduce the state vector $\mathbf{s}_t := (\mathbf{p}_t, \mathbf{q}_t^c)$, which we assume to be a stationary random process. We can then express the loss function $f(\mathbf{p}_t, \mathbf{q}^g - \mathbf{q}_t^c)$ as $f(\mathbf{q}^g; \mathbf{s}_t)$. Substituting this representation into the original problem (8.9), yields

$$\mathbf{q}_t^{g,*}(\mathbf{s}_t) := \arg \min_{\underline{\mathbf{q}}^g \leq \mathbf{q}^g \leq \overline{\mathbf{q}}^g} \mathbb{E}\big[f(\mathbf{q}^g; \mathbf{s}_t)\big]. \tag{8.10}$$

Rather than the unreliable and possibly obsolete instantaneous $\mathbf{q}_t^{g,*}$ found through (8.4), problem (8.10) is expected to yield smoother power control decisions. However, practical implementation faces formi-dable challenges, including the daunting task of computing the expectation in (8.10), even if we were to know the probability density function of $\mathbf{s}_t$. Another hurdle arises from the computational complexity associated with the non-convex constraint (8.5e). To overcome these obstacles and efficiently approximate $\mathbf{q}^{g,*}$, we introduce a statistical learning approach in the following section.

8.3 Statistical Learning

The rapid expansion of renewable energy sources is reshaping the energy landscape, progressively replacing traditional power generation methods. This transformation highlights the increasing need for controllable and flexible resources capable of managing fluctuations in energy consumption and generation. In the following section, we introduce an innovative parameterization approach to tackle the reactive power control problem. Furthermore, we present a learning solver based on DNNs as a promising solution to this evolving challenge.

8.3.1 Parameterization

Rather than pursuing an exact solution for (8.10), let us explore a parameterization approach for reactive power compensation

$$\mathbf{q}^g = \phi(\mathbf{s}; \boldsymbol{\theta}), \tag{8.11}$$

where $\phi(\mathbf{s}; \boldsymbol{\theta})$ is a function represented, potentially as a DNNs, and $\boldsymbol{\theta} \in \mathbb{R}^d$ holds all the relevant unknown parameters.

Elaborating on this concept, the determination of the optimal reactive power control $\mathbf{q}^{g,*}$ in (8.10) essentially revolves around identifying the optimal parameter vector $\boldsymbol{\theta}^*$. This vector should be selected to minimize the expected loss, which is formulated as

$$\boldsymbol{\theta}^* = \arg\min_{\boldsymbol{\theta}} \mathbb{E}\big[f(\phi(\mathbf{s}; \boldsymbol{\theta}); \mathbf{s})\big]. \tag{8.12}$$

To uncover the optimal $\boldsymbol{\theta}^*$, a common strategy involves the application of gradient descent type algorithms. In pursuit of this goal, it becomes imperative to derive the gradient of the objective function in (8.12) concerning $\boldsymbol{\theta}$, i.e., $\nabla_{\boldsymbol{\theta}}\mathbb{E}\big[f(\phi(\mathbf{s}; \boldsymbol{\theta}); \mathbf{s})\big]$. In practical cases, obtaining an analytical expression for $f(\phi(\mathbf{s}; \boldsymbol{\theta}); \mathbf{s})$ in terms of either $\phi(\mathbf{s}; \boldsymbol{\theta})$ or $\boldsymbol{\theta}$ is not feasible. For instance, in (8.5a–8.5f), the loss function f implicitly depends on $\mathbf{q}^g$. Instead, we can observe the function's value, f, for any operational state of the grid $(\mathbf{p}, \mathbf{q}, \mathbf{P}, \mathbf{Q}, \mathbf{l}, v)$[as per (8.5a–8.5f)], which can then be utilized to estimate the gradient. This serves as the motivation for developing a model-free approach.

For a given set of iterations and reactive power realizations, represented as $\{\tilde{\theta}, \tilde{\mathbf{q}}^g\}$, we can directly observe the corresponding loss function values, denoted as $\tilde{f}(\tilde{\mathbf{q}}^g_{\boldsymbol{\theta}}; \mathbf{s})$, from the system. With this information and the values of $\{\tilde{\theta}, \tilde{\mathbf{q}}^g\}$, we can update the parameter vector $\boldsymbol{\theta}$ using the policy gradient method (Sutton et al. 1999). This method constructs a gradient estimate solely based on function observations, enabling effective parameter refinement.

In this context, a control policy is a function mapping from state vectors, denoted as $\mathbf{s}$, to reactive power control decisions, or actions, represented by $\mathbf{q}^g$. First, let's consider a stochastic control policy, denoted as π, which defines a conditional distribution of all possible decisions $\mathbf{q}^g$ given the current state $\mathbf{s}$. We can express the probability of selecting action $\mathbf{q}^g$ at state $\mathbf{s}$ as $\pi_{\mathbf{s},\boldsymbol{\theta}}(\mathbf{q}^g)$. With this notation, we can derive the gradient of the expected value $\mathbb{E}\big[f(\phi(\mathbf{s}; \boldsymbol{\theta}); \mathbf{s})\big]$ with respect to the parameter vector $\boldsymbol{\theta}$ as follows

$$\nabla_{\boldsymbol{\theta}}\mathbb{E}\big[f(\phi(\mathbf{s}; \boldsymbol{\theta}); \mathbf{s})\big] = \nabla_{\boldsymbol{\theta}} \int_{\mathbf{s}} f(\phi(\mathbf{s}; \boldsymbol{\theta}); \mathbf{s}) \Pr(\mathbf{s}) ds \tag{8.13a}$$

$$= \nabla_{\boldsymbol{\theta}} \int_{\mathbf{s}} \int_{\mathbf{q}^g} f(\mathbf{q}^g; \mathbf{s}) \pi_{\mathbf{s},\boldsymbol{\theta}}(\mathbf{q}^g) \Pr(\mathbf{s}) d\mathbf{q}^g ds \tag{8.13b}$$

$$= \int_{\mathbf{s}} \int_{\mathbf{q}^g} f(\mathbf{q}^g; \mathbf{s}) \frac{\nabla_{\boldsymbol{\theta}}\pi_{\mathbf{s},\boldsymbol{\theta}}(\mathbf{q}^g)}{\pi_{\mathbf{s},\boldsymbol{\theta}}(\mathbf{q}^g)} \pi_{\mathbf{s},\boldsymbol{\theta}}(\mathbf{q}^g) \Pr(\mathbf{s}) d\mathbf{q}^g ds \tag{8.13c}$$

$$= \mathbb{E}_{\mathbf{q}^g,\mathbf{s}}\big[f(\mathbf{q}^g; \mathbf{s}) \nabla_{\boldsymbol{\theta}} \log \pi_{\mathbf{s},\boldsymbol{\theta}}(\mathbf{q}^g)\big]. \tag{8.13d}$$

In this equation, $\Pr(\mathbf{s})$ represents the probability distribution of state $\mathbf{s}$, and $\mathbf{q}^g$ is drawn from the conditional distribution $\pi_{\mathbf{s},\boldsymbol{\theta}}(\cdot)$. We can simplify the computation of $\nabla_{\boldsymbol{\theta}}\mathbb{E}\big[f(\phi(\mathbf{s};\boldsymbol{\theta});\mathbf{s})\big]$ by expressing it as the expectation of the function $f(\mathbf{q}^g;\mathbf{s})$ multiplied by the gradient of the policy distribution $\nabla_{\boldsymbol{\theta}}\log\pi_{\mathbf{s},\boldsymbol{\theta}}(\mathbf{q}^g)$. This approach is particularly valuable when we have an analytical expression for $\pi_{\mathbf{s},\boldsymbol{\theta}}(\mathbf{q}^g)$. In cases where we have access to an analytical form of $\pi_{\mathbf{s},\boldsymbol{\theta}}(\mathbf{q}^g)$, we can further approximate the expectation on the right-hand side of (8.13a–8.13d) with a sample mean. This leads to the following gradient estimate based on previous function observations

$$\widehat{\nabla_{\boldsymbol{\theta}}\mathbb{E}}\big[f(\phi(\mathbf{s};\boldsymbol{\theta});\mathbf{s})\big] = \hat{f}(\hat{\mathbf{q}}^g_{\boldsymbol{\theta}};\mathbf{s})\nabla_{\boldsymbol{\theta}}\log\pi_{\mathbf{s},\boldsymbol{\theta}}(\hat{\mathbf{q}}^g_{\boldsymbol{\theta}}), \tag{8.14}$$

where $\hat{\mathbf{q}}^g_{\boldsymbol{\theta}}$ represents the injected reactive power into the distribution grid, sampled from the distribution $\pi_{\mathbf{s},\boldsymbol{\theta}}(\cdot)$, and $\hat{f}(\hat{\mathbf{q}}^g_{\boldsymbol{\theta}};\mathbf{s})$ denotes the corresponding loss function value obtained by solving (8.5a–8.5f).

Previously, it was assumed that the policy $\phi(\mathbf{s};\boldsymbol{\theta})$ is stochastic. In deterministic cases, where the distribution becomes a delta function, i.e., $\pi_{\mathbf{s},\boldsymbol{\theta}}(\mathbf{x}) = \delta(\mathbf{x}-\phi(\mathbf{s};\boldsymbol{\theta}))$. To compute $\nabla_{\boldsymbol{\theta}}\log\pi_{\mathbf{s},\boldsymbol{\theta}}(\hat{\mathbf{q}}^g_{\boldsymbol{\theta}})$ in (8.14), one can approximate the delta function with a known density function centered around $\phi(\mathbf{s};\boldsymbol{\theta})$. To account for the power constraint $\underline{\mathbf{q}}^g \leq \mathbf{q}^g \leq \overline{\mathbf{q}}^g$, we consider a truncated Gaussian distribution with fixed support on the domain $[\underline{\mathbf{q}}^g, \overline{\mathbf{q}}^g]$, which will be discussed in the following subsection.

8.3.2 Model-Free Learning

To determine the policy $\pi_{\mathbf{s},\boldsymbol{\theta}}(\cdot)$, we confine our exploration to the ever-popular realm of parameterization, specifically DNNs (Sze et al. 2017). DNNs have, in recent times, exhibited exceptional capabilities across a multitude of domains such as computer vision, speech recognition, and robotics. A deep neural network can effectively tackle the 'curse of dimensionality' by extracting low-dimensional representation for high-dimensional data (Sze et al. 2017).

Consider a feed-forward DNNs intricately connected to a truncated Gaussian probability density function $\pi_{\mathbf{s},\boldsymbol{\theta}}(\cdot)$ module, as depicted in Fig. 8.2. This network takes the state vector $\mathbf{s}$ as input, followed by a sequence of L fully connected hidden layers, each employing ReLU activation functions. At its output, the DNNs generates a set of mean and standard deviation pairs, denoted as $\{\mu_m, \delta_m\}_{m=1}^M$, where each pair corresponds to one of the M truncated Gaussian distributions. By forwarding these network outputs into the probability density function $\pi_{\mathbf{s},\boldsymbol{\theta}}(\cdot)$ module, we can sample the reactive power compensation vector $\mathbf{q}^g$ from $\pi_{\mathbf{s},\boldsymbol{\theta}}(\mathbf{q}^g)$. Collecting all the weights of the DNNs into the parameter vector $\boldsymbol{\theta}$, we obtain a function approximation for estimating the reactive power compensation, i.e., $\mathbf{q}^g(\mathbf{s}) = \phi(\mathbf{s};\boldsymbol{\theta})$.

Utilizing the gradient estimate in Eq. (8.14), we can systematically update the weights $\boldsymbol{\theta}$ through the following iterative process

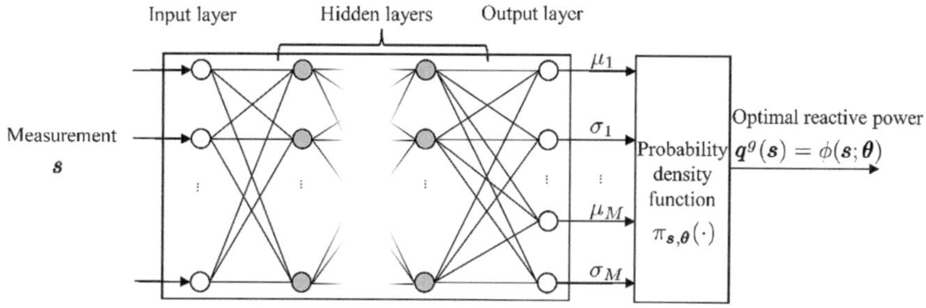

Fig. 8.2 Statistical learning architecture

$$\boldsymbol{\theta}_{t+1} = \boldsymbol{\theta}_t - \beta_t \widehat{\nabla_{\boldsymbol{\theta}} \mathbb{E}}\big[f(\phi(\mathbf{s}; \boldsymbol{\theta}); \mathbf{s})\big]\big|_{\boldsymbol{\theta}=\boldsymbol{\theta}_t}, \tag{8.15}$$

where $\beta_t > 0$ is a preselected learning rate. This update in (7.16) is a model-free approach, since it does not require explicit knowledge about the actual form of the function $f(\cdot)$ or distribution of **s**.This method is distinct from traditional supervised approaches that rely on training labeled data (Zhang et al. 2019b). The developed approach is unsupervised, eliminating the requirement for labeled data and enabling direct solution of Eq. (8.10).

Algorithm 8.1 A deep policy gradient approach to reactive power control

Training phase:
1: **Initialize:** $\boldsymbol{\theta}$
2: **for** $t = 1, 2, \ldots, T$ **do**
3: Observe historical measurement $\mathbf{s}_t$
4: Feed $\mathbf{s}_t$ into the DNNs
5: Obtain DNNs output mean $\boldsymbol{\mu}_t$ and variance $\boldsymbol{\sigma}_t$
6: Feed $\boldsymbol{\mu}_t$ and $\boldsymbol{\sigma}_t$ into $\pi_{\mathbf{s}_t,\boldsymbol{\theta}_t}(\cdot)$
7: Draw a sample $\hat{\mathbf{q}}^g_{\boldsymbol{\theta}}$ from the distribution $\pi_{\mathbf{s}_t,\boldsymbol{\theta}_t}(\mathbf{q}^g)$
8: Obtain an estimate for $\hat{f}(\mathbf{s}, \hat{\mathbf{q}}^g_{\boldsymbol{\theta}_t})$ via (8.7a–8.7c)
9: Calculate $\widehat{\nabla_{\boldsymbol{\theta}} \mathbb{E}}\big[f(\mathbf{s}_t, \phi(\mathbf{s}_t; \boldsymbol{\theta}_t))\big]$ via (8.14)
10: Update $\boldsymbol{\theta}_{t+1}$ according to (7.16)
11: **end for**

Inference phase:
1: **for** $\tau = 1, 2, \ldots$ **do**
2: Feed real-time measurement $\mathbf{s}_\tau$ into the trained DNNs
3: Obtain the DNNs output mean $\boldsymbol{\mu}_t$ and variance $\boldsymbol{\sigma}_t$
4: Feed $\boldsymbol{\mu}_t$ and $\boldsymbol{\sigma}_t$ into $\pi_{\mathbf{s}_\tau,\boldsymbol{\theta}_\tau}(\cdot)$
5: Draw a sample $\hat{\mathbf{q}}^g_{\boldsymbol{\theta}_\tau}$ from the distribution $\pi_{\mathbf{s}_\tau,\boldsymbol{\theta}_\tau}(\mathbf{q}^g)$
6: **end for**

The reactive power control procedure, as outlined in Algorithm 8.1, unfolds in two distinct phases: offline training and online inference phases, as depicted in Fig. 8.3.

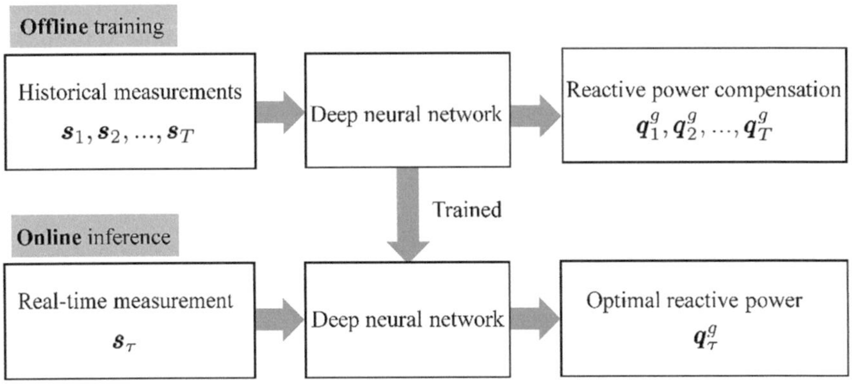

Fig. 8.3 Two-phase reactive power control procedure

During the training phase, historical or simulated datum $\mathbf{s}$ is input into the DNNs. For a given input datum $\mathbf{s}_t$, our network generates a reactive power compensation $\mathbf{q}_t^g = \phi(\mathbf{s}_t; \boldsymbol{\theta}_t)$. Subsequently, the distribution network provides a loss value for this state-action pair $(\mathbf{s}_t, \mathbf{q}_t^g)$, which can also be obtained by solving Eq. (8.5a–8.5f). Finally, a gradient estimate is computed using the policy gradient method detailed in Eq. (8.14). This estimate informs the update of the NNs' weight parameters in accordance with Eq. (7.16). The trained DNNs becomes invaluable during the inference phase. By taking the real-time state vector $\mathbf{s}_t = (\mathbf{p}_t, \mathbf{q}_t^c)$ as input, the trained DNNs furnishes the optimal reactive power compensation $\mathbf{q}_t^g$ to be implemented in the grid. Note that the proposed statistical learning approach is desirable for real-time reactive power control, as it shifts the computational burden of tackling non-convex optimization to offline training of a neural network.

8.4 Numerical Tests

In this section, we turn our attention to the empirical evaluation of our proposed statistical learning scheme, conducted on a real-world 47-bus feeder renowned for its high penetration of renewable energy sources (Farivar et al. 2011). A visual representation of this distribution grid can be observed in Fig. 8.4: Schematic diagram of the 47-bus industrial distribution feeder. Bus 1 is the substation, and the 6 loads connected to it model other feeders on this substation. Within this grid, $M = 5$ smart inverters are strategically positioned on buses 2, 16, 18, 21, and 22, each boasting different power capacities, specifically 30, 80, 300, 400, and 200 kW, respectively. It's pertinent to note that a uniform power factor of 0.8 has been assumed for all loads within this system.

The dataset utilized for both training and testing was sourced from the Smart * project, with a division made between consumption and solar generation data. These

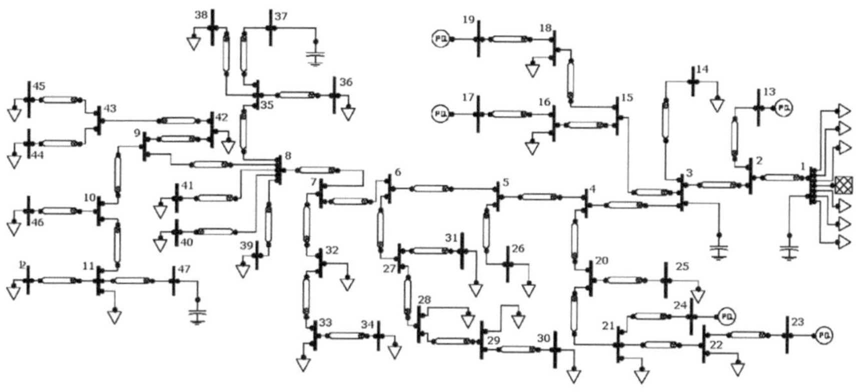

Fig. 8.4 Schematic diagram of the 47-bus distribution feeder

records were specifically collected on August 24, 2011, as documented in (Barker et al. 2012). In our simulations, we harnessed the computational power of the CVX toolbox (Grant and Boyd 2014) to effectively solve the SOCP problem presented in (8.7a–8.7c). This facilitated the evaluation of $\hat{f}(\hat{\mathbf{q}}_{\boldsymbol{\theta}}^{g}; \mathbf{s})$. The architecture of the employed DNNs was characterized by three fully connected hidden layers, featuring 48, 32, and 16 neurons per respective layer. Our computational framework relied on 'TensorFlow' (Abadi et al. 2016), efficiently executed on an NVIDIA Titan X GPU boasting 12 GB of RAM. The DNNs' weight parameters underwent updates using the back-propagation algorithm, with optimization carried out via the 'Adam' optimizer. Throughout our experiments, we maintained a fixed learning rate of 0.01, and the batch size consistently held at 30 samples over the course of 40 testing epochs.

To gauge the effectiveness of our proposed approach, we established a baseline for comparison. In this baseline case, we assumed perfect observations of active and reactive power injections, denoted as $(\mathbf{p}_t, \mathbf{q}_t^c)$, at the outset of per time slot t. Under these ideal conditions, the optimal reactive power control can be determined by solving the following optimization problem

$$f(\mathbf{p}_t, \mathbf{q}_t^g - \mathbf{q}_t^c) = \min_{\substack{\mathbf{q}_t^g, \mathbf{P}_t, \mathbf{Q}_t \\ \mathbf{l}_t, \mathbf{v}_t}} \sum_{n=1}^{L} r_{n,t} \ell_{n,t}, \tag{8.16a}$$

subject to (8.7b–8.7c)

where $\mathbf{q}_t^g$ is treated as an optimization variable. It is worth highlighting that addressing this problem in real-time poses significant computational challenges, especially for large-scale systems. In contrast, our proposed approach computes $\mathbf{q}_t^g$ with just a few matrix–vector multiplications. In Fig. 8.5, the red curve represents the observed loss for our proposed approach, while the blue curve illustrates the loss for the deterministic optimal solution obtained via (8.16a–8.16b) during the training

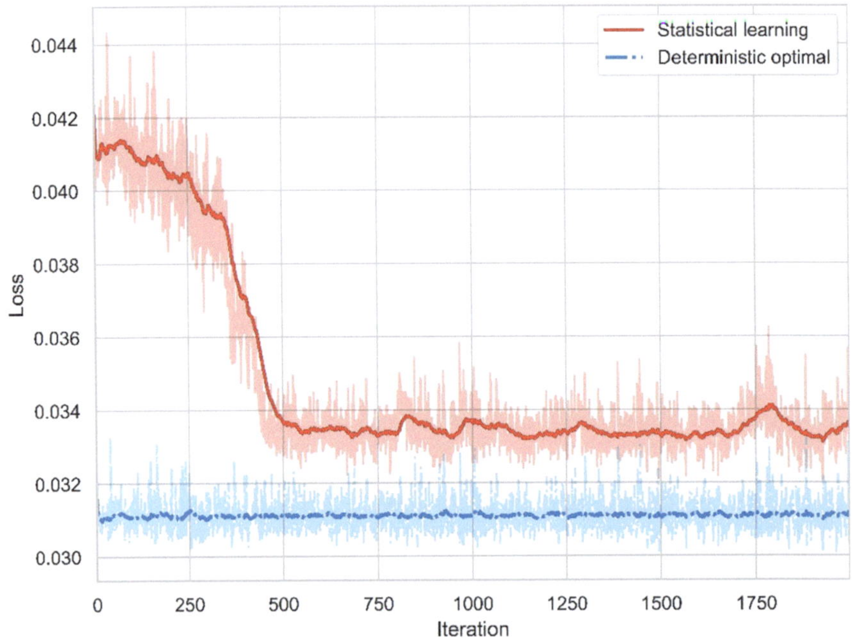

Fig. 8.5 The training loss of the statistical learning approach compared with the baseline (optimal)

phase. Light-colored curves correspond to the actual observed losses, whereas dark-colored curves depict running averages. Evidently, our model-free approach effectively learns to make optimal decisions for $\mathbf{q}_t^g$. In the inference phase, the loss of our proposed approach compared to the baseline is depicted in Fig. 8.6. This plot clearly demonstrates that our model-free approach achieves near-optimal reactive power control decisions. Moreover, the running time of our proposed approach is approximately one order of magnitude shorter than that of the optimization-based approach.

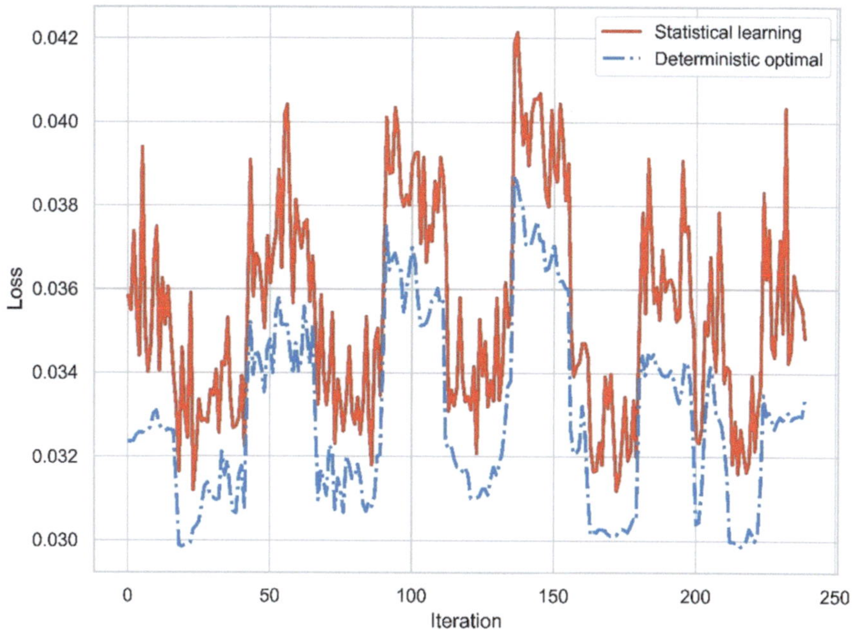

Fig. 8.6 The inference loss values of statistical learning approach and the baseline (optimal)

Chapter 9
Conclusion

As it was manifested, the implementation of accurate SE and making robust decision for power systems was a worthwhile and in-depth investigation. Leveraging the advancements in artificial intelligence technology, this monograph introduced an innovative framework for monitoring and controlling power grids. This framework adeptly integrated, in its initial stage, unique NNs specifically designed for interpreting power grid topology and predicting future states. Additionally, the principles of composite optimization theory were incorporated; it simplified the complexity inherent in the estimation of power system states, facilitating rapid and accurate assessments of system conditions. Moreover, this framework took a step further by intricately interweaving a data-driven approach with physics-based optimization, thereby unleashing the full potential of system flexibility. Ultimately, by employing NNs to parameterize the relationship between system states and control policies, that framework excelled in providing precise and targeted robust control strategies when confronted with newly estimated system states. The numerical experiments in each chapter validated our theoretical findings and demonstrated the effectiveness, robustness, and computational efficiency of our approach.

G. Wang et al., *Data-driven Optimization and Control for Autonomous Energy Systems*,
https://doi.org/10.1007/978-981-95-1782-4_9

References

Abadi M, Agarwal A, Barham P, Brevdo E, Chen Z, Citro C, Corrado GS, Davis A, Dean J, Devin M et al (2016) Tensorflow: large-scale machine learning on heterogeneous distributed systems. arXiv preprint arXiv:1603.04467

Abur A, Celik MK (1991) A fast algorithm for the weighted least absolute value state estimation (for power systems). IEEE Trans Power Syst 6(1):1–8

Abur A, Exposito AG (2004) Power system state estimation: theory and implementation. CRC Press

Adeodato PJ, Braga PL, Arnaud AL, Vasconcelos GC, Guedes F, Menezes HB, Limeira GO (2010) Domain driven data mining for unavailability estimation of electrical power grids. In: Trends in applied intelligent systems: 23rd international conference on industrial engineering and other applications of applied intelligent systems, IEA/AIE 2010, Cordoba, Spain, June 1–4, 2010, Proceedings, Part II 23. Springer, pp 357–366

Agalgaonkar YP, Pal BC, Jabr RA (2015) Stochastic distribution system operation considering voltage regulation risks in the presence of PV generation. IEEE Trans Sustain Energy 6(4):1315–1324

Aggarwal HK, Mani MP, Jacob M (2018) MoDL: model-based deep learning architecture for inverse problems. IEEE Trans Med Imaging 38(2):394–405

Babaeinejadsarookolaee S, Birchfield A, Christie RD, Coffrin C, DeMarco C, Diao R, Ferris M, Fliscounakis S, Greene S, Huang R et al (2019) The power grid library for benchmarking ac optimal power flow algorithms. arXiv preprint arXiv:1908.02788

Balan R (2016) Reconstruction of signals from magnitudes of redundant representations: the complex case. Found Comput Math 16:677–721

Bandi C, Bertsimas D (2014) Robust option pricing. Eur J Oper Res 239(3):842–853

Baran ME, El-Markabi IM (2007) A multiagent-based dispatching scheme for distributed generators for voltage support on distribution feeders. IEEE Trans Power Syst 22(1):52–59

Baran M, Wu FF (1989a) Optimal sizing of capacitors placed on a radial distribution system. IEEE Trans Power Delivery 4(1):735–743

Baran ME, Wu FF (1989b) Network reconfiguration in distribution systems for loss reduction and load balancing. IEEE Trans Power Delivery 4(2):1401–1407

Baran ME, Wu FF (1989c) Optimal capacitor placement on radial distribution systems. IEEE Trans Power Delivery 4(1):725–734

Barbeiro PP, Krstulovic J, Teixeira H, Pereira J, Soares FJ, Iria JP (2014) State estimation in distribution smart grids using autoencoders. In: 2014 IEEE 8th international power engineering and optimization conference (PEOCO2014). IEEE, pp 358–363

Barker S, Mishra A, Irwin D, Cecchet E, Shenoy P, Albrecht J et al (2012) Smart*: an open data set and tools for enabling research in sustainable homes. SustKDD 111(112):108

Bazrafshan M, Gatsis N (2016) Decentralized stochastic optimal power flow in radial networks with distributed generation. IEEE Trans Smart Grid 8(2):787–801

G. Wang et al., *Data-driven Optimization and Control for Autonomous Energy Systems*,
https://doi.org/10.1007/978-981-95-1782-4

Ben-Tal A, Nemirovski A (2001) Lectures on modern convex optimization: analysis, algorithms, and engineering applications. SIAM

Bertsekas DP (1997) Nonlinear programming. J Oper Res Soc 48(3):334–334

Bertsekas DP (1999) Nonlinear programming/Dimitri P. Bertsekas, Athena Scientific, Belmont (Mass.), 2nd edn

Bhela S, Kekatos V, Veeramachaneni S (2017) Enhancing observability in distribution grids using smart meter data. IEEE Trans Smart Grid 9(6):5953–5961

Blackburn JL, Domin TJ (2006) Protective relaying: principles and applications. CRC Press

Blanchet J, Murthy K (2019) Quantifying distributional model risk via optimal transport. Math Oper Res 44(2):565–600

Boyd S, Parikh N, Chu E, Peleato B, Eckstein J et al (2011) Distributed optimization and statistical learning via the alternating direction method of multipliers. Found Trends® Mach Learn 3(1):1–122

Brandwood D (1983) A complex gradient operator and its application in adaptive array theory. In: IEE Proceedings H (Microwaves, Optics and Antennas), vol 130, no 1. IET Digital Library, pp 11–16

Burke JV (1985) Descent methods for composite nondifferentiable optimization problems. Math Program 33:260–279

Burke JV, Ferris MC (1995) A Gauss—Newton method for convex composite optimization. Math Program 71(2):179–194

Cao D, Zhao J, Hu W, Ding F, Huang Q, Chen Z, Blaabjerg F (2020) Model-free voltage regulation of unbalanced distribution network based on surrogate model and deep reinforcement learning. arXiv preprint arXiv:2006.13992

Caro E, Conejo A (2012) State estimation via mathematical programming: a comparison of different estimation algorithms. IET Gener Transm Distrib 6(6):545

Carta JA, Ramirez P, Velazquez S (2009) A review of wind speed probability distributions used in wind energy analysis: case studies in the Canary Islands. Renew Sustain Energy Rev 13(5):933–955

Carvalho PM, Correia PF, Ferreira LA (2008) Distributed reactive power generation control for voltage rise mitigation in distribution networks. IEEE Trans Power Syst 23(2):766–772

Christie R (2000) Power systems test case archive, University of Washington. Electrical Engineering. https://www2.ee.washington.edu/research/pstca

Claessens BJ, Vrancx P, Ruelens F (2016) Convolutional neural networks for automatic state-time feature extraction in reinforcement learning applied to residential load control. IEEE Trans Smart Grid 9(4):3259–3269

Conejo AJ, Carrión M, Morales JM et al (2010) Decision making under uncertainty in electricity markets, vol 1. Springer

Constante-Flores GE, Conejo AJ, Wang J (2022) Stealthy monitoring-control attacks to disrupt power system operations. Electric Power Syst Res 203:107636

Couture T, Cory K (2009) State clean energy policies analysis (SCEPA) project: an analysis of renewable energy feed-in tariffs in the United States (revised). Tech. Rep. National Renewable Energy Lab. (NREL), Golden, CO (United States)

Da Silva AL, Do Coutto Filho M, De Queiroz J (1983) State forecasting in electric power systems. In: IEE Proceedings C (Generation, Transmission and Distribution), vol 130, no 5. IET, pp 237–244

Dall'Anese E, Dhople SV, Giannakis GB (2014) Optimal dispatch of photovoltaic inverters in residential distribution systems. IEEE Trans Sustain Energy 5(2):487–497

Davis D, Drusvyatskiy D (2019) Stochastic model-based minimization of weakly convex functions. SIAM J Optim 29(1):207–239

Debs AS, Larson RE (1970) A dynamic estimator for tracking the state of a power system. IEEE Trans Power Appar Syst 7:1670–1678

Diao R, Wang Z, Shi D, Chang Q, Duan J, Zhang X (2019) Autonomous voltage control for grid operation using deep reinforcement learning. In: 2019 IEEE power and energy society general meeting (PESGM). IEEE, pp 1–5

Do Coutto Filho MB, de Souza JCS (2009) Forecasting-aided state estimation—Part I: Panorama. IEEE Trans Power Syst 24(4):1667–1677

Do Coutto Filho MB, de Souza JCS, Freund RS (2009) Forecasting-aided state estimation—Part II: implementation. IEEE Trans Power Syst 24(4):1678–1685

Duan J, Xu H, Liu W (2017) Q-learning-based damping control of wide-area power systems under cyber uncertainties. IEEE Trans Smart Grid 9(6):6408–6418

Duchi J, Hazan E, Singer Y (2011) Adaptive subgradient methods for online learning and stochastic optimization. J Mach Learn Res 12(7)

Duchi JC, Ruan F (2018) Stochastic methods for composite and weakly convex optimization problems. SIAM J Optim 28(4):3229–3259

Duchi JC, Ruan F (2019) Solving (most) of a set of quadratic equalities: composite optimization for robust phase retrieval. Inf Inference: J IMA 8(3):471–529

Durgaprasad G, Thakur S (1998) Robust dynamic state estimation of power systems based on M-estimation and realistic modeling of system dynamics. IEEE Trans Power Syst 13(4):1331–1336

Džafić I, Jabr RA, Hrnjić T (2018) Hybrid state estimation in complex variables. IEEE Trans Power Syst 33(5):5288–5296

Ernst D, Glavic M, Wehenkel L (2004) Power systems stability control: reinforcement learning framework. IEEE Trans Power Syst 19(1):427–435

Farivar M, Clarke CR, Low SH, Chandy KM (2011) Inverter VAR control for distribution systems with renewables. In: 2011 IEEE international conference on smart grid communications (SmartGridComm). IEEE, pp 457–462

Farivar M, Low SH (2013) Branch flow model: relaxations and convexification—Part I. IEEE Trans Power Syst 28(3):2554–2564

Feeders DT (2011) IEEE PES distribution system analysis subcommittee. Online Available: http://www.ewh.ieee.org/soc/pes/dsacom/testfeeders/index.html

Gan L, Li N, Topcu U, Low SH (2014) Exact convex relaxation of optimal power flow in radial networks. IEEE Trans Autom Control 60(1):72–87

Gandhi MA, Mili L (2009) Robust Kalman filter based on a generalized maximum-likelihood-type estimator. IEEE Trans Signal Process 58(5):2509 2520

Gatsis N, Marques AG (2014) A stochastic approximation approach to load shedding inpower networks. In: 2014 IEEE international conference on acoustics, speech and signal processing (ICASSP). IEEE, pp 6464–6468

Gatsis N, Ribeiro A, Giannakis GB (2010) A class of convergent algorithms for resource allocation in wireless fading networks. IEEE Trans Wireless Commun 9(5):1808–1823

Giannakis GB, Kekatos V, Gatsis N, Kim S-J, Zhu H, Wollenberg BF (2013) Monitoring and optimization for power grids: a signal processing perspective. IEEE Signal Process Mag 30(5):107–128

Göl M, Abur A (2014) LAV based robust state estimation for systems measured by PMUs. IEEE Trans Smart Grid 5(4):1808–1814

Goodfellow I, Bengio Y, Courville A (2016) Deep learning. MIT press

Grant M, Boyd S (2014) CVX: Matlab software for disciplined convex programming, version 2.1

Grant M, Boyd S, Ye Y (2009) CVX: Matlab software for disciplined convex programming

Gregor K, LeCun Y (2010) Learning fast approximations of sparse coding. In: Proceedings of the 27th international conference on international conference on machine learning, pp 399–406

Hassanzadeh M, Evrenosoğlu CY (2012) Power system state forecasting using regression analysis. In: 2012 IEEE power and energy society general meeting. IEEE, pp 1–6

Hassanzadeh M, Evrenosoğlu CY, Mili L (2015) A shortterm nodal voltage phasor forecasting method using temporal and spatial correlation. IEEE Trans Power Syst 31(5):3881–3890

He K, Zhang X, Ren S, Sun J (2016a) Deep residual learning for image recognition. In: Proceedings of the IEEE conference on computer vision and pattern recognition, pp 770–778

He K, Zhang X, Ren S, Sun J (2016b) Identity mappings in deep residual networks. In: Computer vision–ECCV 2016: 14th European Conference, Amsterdam, The Netherlands, October 11–14, 2016, Proceedings, Part IV 14. Springer, pp 630–645
Hering M, Conca A, Edidin D, Vinzant C (2013) An algebraic characterisation of injectivity in phase retrieval
Hu X, Hu H, Verma S, Zhang Z-L (2020) Physics-guided deep neural networks for power flow analysis. IEEE Trans Power Syst 36(3):2082–2092
Huang G, Ge C, Xiong T, Song S, Yang L, Liu B, Yin W, Wu C (2021) Large scale air pollution prediction with deep convolutional networks. Sci China Inf Sci 64:1–11
Huber PJ (2004) Robust statistics, vol 523. Wiley
Jabr R, Pal B (2003) Iteratively re-weighted least absolute value method for state estimation. IEE Proc-Gener Transm Distrib 150(4):385–391
Jabr R, P B (2004) Iteratively reweighted least-squares implementation of the WLAV state-estimation method. IEE Proc Gener Transm Distrib 151(1):103–108
Jalali M, Kekatos V, Gatsis N, Deka D (2019) Designing reactive power control rules for smart inverters using support vector machines. IEEE Trans Smart Grid 11(2):1759–1770
Jin X, Edwards RM, Ray A (2010) Feature extraction for data-driven fault detection in nuclear power plants. Trans Am Nucl Soc 102(163):163–164
Kekatos V, Giannakis GB (2012) Distributed robust power system state estimation. IEEE Trans Power Syst 28(2):1617–1626
Kekatos V, Wang G, Conejo AJ, Giannakis GB (2014) Stochastic reactive power management in microgrids with renewables. IEEE Trans Power Syst 30(6):3386–3395
Kekatos V, Wang G, Zhu H, Giannakis GB (2017) PSSE redux: convex relaxation, decentralized, robust, and dynamic approaches. arXiv preprint arXiv:1708.03981
Kekatos V, Zhang L, Giannakis GB, Baldick R (2015) Voltage regulation algorithms for multiphase power distribution grids. IEEE Trans Power Syst 31(5):3913–3923
Kersting WH (2018) Distribution system modeling and analysis. In: Electric power generation, transmission, and distribution. CRC press, pp 26–1
Kingma DP, Ba J (2014) Adam: a method for stochastic optimization. arXiv preprint arXiv:1412.6980
Kipf TN, Welling M (2016) Semi-supervised classification with graph convolutional networks. arXiv preprint arXiv:1609.02907
Kotiuga WW, Vidyasagar M (1982) Bad data rejection properties of weughted least absolute value techniques applied to static state estimation. IEEE Trans Power Appar Syst 4:844–853
Koutsopoulos I, Tassiulas L (2011) Control and optimization meet the smart power grid: Scheduling of power demands for optimal energy management. In: Proceedings of the 2nd international conference on energy-efficient computing and networking, pp 41–50
Kruger M, Ding SX, Haghani A, Engel P, Jeinsch T (2013) A data-driven approach for sensor fault diagnosis in gearbox of wind energy conversion system. In: 2013 10th IEEE international conference on control and automation (ICCA). IEEE, pp 227–232
Kurakin A, Goodfellow IJ, Bengio S (2018) Adversarial examples in the physical world. In: Artificial intelligence safety and security. Chapman and Hall/CRC, pp 99–112
Kusiak A, Zhang Z (2010) Short-horizon prediction of wind power: a data-driven approach. IEEE Trans Energy Convers 25(4):1112–1122
LeCun Y, Bengio Y, Hinton G (2015) Deep learning. Nature 521(7553):436–444
Lee W, Kim M, Cho D-H (2018) Deep power control: transmit power control scheme based on convolutional neural network. IEEE Commun Lett 22(6):1276–1279
Lewis AS, Wright SJ (2016) A proximal method for composite minimization. Math Program 158:501–546
Li P, Li N, Li QM, Cao M, Chen HX (2011) Prediction model for power transmission line icing load based on data-driven. Adv Mater Res 143:1295–1299

Liao Y, Weng Y, Wu M, Rajagopal R (2015) Distribution grid topology reconstruction: an information theoretic approach. In: 2015 North American power symposium (NAPS). IEEE, pp 1–6

Lin H, Jegelka A (2018) Resnet with one-neuron hidden layers is a universal approximator. Adv Neural Inf Process Syst 31

Lin W, Thomas R, Bitar E (2018) Real-time voltage regulation in distribution systems via decentralized PV inverter control

Lingala SG, Jacob M (2013) Blind compressive sensing dynamic MRI. IEEE Trans Med Imaging 32(6):1132–1145

Liu E, Bebic J (2008) Distribution system voltage performance analysis for high-penetration photovoltaics. Tech Rep. National Renewable Energy Lab. (NREL), Golden, CO (United States)

Low SH (2014) Convex relaxation of optimal power flow—Part II: exactness. IEEE Trans Control Netw Syst 1(2):177–189

Lu R, Hong SH, Yu M (2019) Demand response for home energy management using reinforcement learning and artificial neural network. IEEE TransSmart Grid 10(6):6629–6639

Manitsas E, Singh R, Pal BC, Strbac G (2012) Distribution system state estimation using an artificial neural network approach for pseudo measurement modeling. IEEE Trans Power Syst 27(4):1888–1896

Markouchevitch AI (1977) Theory of functions of a complex variable, vol 1. American Mathematical Soc

May E (2013) Feed-in tariff: a policy tool encouraging deployment of renewable electricity technologies. Accessed August

Menke JH, Bornhorst N, Braun M (2019) Distribution system monitoring for smart power grids with distributed generation using artificial neural networks. Int J Electr Power Energy Syst 113:472–480

Merrill HM, Schweppe FC (1971) Bad data suppression in power system static state estimation. IEEE Trans Power Appar Syst 6:2718–2725

Mestav KR, Luengo-Rozas J, Tong L (2018) State estimation for unobservable distribution systems via deep neural networks. In: 2018 IEEE power and energy society general meeting (PESGM). IEEE, pp 1–5

Mili L, Cheniae M, Vichare N, Rousseeuw P (1996a) Robustification of the least absolute value estimator by means of projection statistics [power system state estimation]. IEEE Trans Power Syst 11(1):216–225

Mili L, Cheniae M, Vichare N, Rousseeuw PJ (1996b) Robust state estimation based on projection statistics [of power systems]. IEEE Trans Power Syst 11(2):1118–1127

Mili L, Cheniae MG, Rousseeuw PJ (1994) Robust state estimation of electric power systems. IEEE Trans Circuits Syst i: Fundam Theory Appl 41(5):349–358

Mnih V, Kavukcuoglu K, Silver D, Rusu AA, Veness J, Bellemare MG, Graves A, Riedmiller M, Fidjeland AK, Ostrovski G et al (2015) Human-level control through deep reinforcement learning. Nature 518(7540):529–533

Morvaj Z (1985) A mathematical model of an electric power system for dynamic state estimation. Electric Power Syst Res 8(3):207–217

Nishiya K, Hasegawa J, Koike T (1982) Dynamic state estimation including anomaly detection and identification for power systems. In: IEEE proceedings C (generation, transmission and distribution), vol 129, no 5. IET, pp 192–198

Nogales FJ, Conejo AJ (2006) Electricity price forecasting through transfer function models. J Oper Res Soc 57(4):350–356

Ostrometzky J, Berestizshevsky K, Bernstein A, Zussman G (2019) Physics-informed deep neural network method for limited observability state estimation. arXiv preprint arXiv:1910.06401

Pan X, Zhao T, Chen M, Zhang S (2020) Deepopf: a deep neural network approach for security-constrained dc optimal power flow. IEEE Trans Power Syst 36(3):1725–1735

Pardalos PM, Vavasis SA (1991) Quadratic programming with one negative eigenvalue is NP-hard. J Global Optim 1(1):15–22

Parikh N, Boyd S et al (2014) Proximal algorithms. Found Trends® Optim 1(3):127–239
Pascanu R, Gulcehre C, Cho K, Bengio Y (2013) How to construct deep recurrent neural networks. arXiv preprint arXiv:1312.6026
Pires RC, Costa AS, Mili L (1999) Iteratively reweighted least-squares state estimation through givens rotations. IEEE Trans Power Syst 14(4):1499–1507
Rathor SK, Saxena D (2020) Energy management system for smart grid: an overview and key issues. Int J Energy Res 44(6):4067–4109
Redondo NJ, Conejo A (1999) Short-term hydro-thermal coordination by Lagrangian relaxation: solution of the dual problem. IEEE Trans Power Syst 14(1):89–95
Ribeiro A (2010) Ergodic stochastic optimization algorithms for wireless communication and networking. IEEE Trans Signal Process 58(12):6369–6386
Riffonneau Y, Bacha S, Barruel F, Ploix S (2011) Optimal power flow management for grid connected PV systems with batteries. IEEE Trans Sustain Energy 2(3):309–320
Rudin LI, Osher S, Fatemi E (1992) Nonlinear total variation based noise removal algorithms. Physica D 60(1–4):259–268
Rumelhart DE, Hinton GE, Williams RJ (1986) Learning representations by back-propagating errors. Nature 323(6088):533–536
Sadeghi A, Wang G, Giannakis GB (2019) Deep reinforcement learning for adaptive caching in hierarchical content delivery networks. IEEE Trans Cogn Commun Netw 5(4):1024–1033
Salehi V, Mohamed A, Mazloomzadeh A, Mohammed OA (2012) Laboratory-based smart power system, part II: control, monitoring, and protection. IEEE Trans Smart Grid 3(3):1405–1417
Schauder C (2014) Advanced inverter technology for high penetration levels of PV generation in distribution systems. Tech Rep. National Renewable Energy Lab. (NREL), Golden, CO (United States)
Schlemper J, Caballero J, Hajnal JV, Price A, Rueckert D (2017) A deep cascade of convolutional neural networks for MR image reconstruction. In: Information processing in medical imaging: 25th international conference, IPMI 2017, Boone, NC, USA, June 25–30, 2017, Proceedings 25. Springer, pp 647–658
Schweppe FC, Wildes J (1970) Power system static-state estimation, Part I: exact model. IEEE Trans Power Appar Syst 1:120–125
Singh MK, Kekatos V, Giannakis GB (2021) Learning to solve the AC-OPF using sensitivity-informed deep neural networks. IEEE Trans Power Syst 37(4):2833–2846
Sinha A, Namkoong H, Volpi R, Duchi J (2017) Certifying some distributional robustness with principled adversarial training. arXiv preprint arXiv:1710.10571
Standard A (1995) C84. 1-1995 electric power systems and equipment voltage ratings (60 Hz)
Standard E (2004) Voltage characteristics of public distribution systems. In: Voltage characteristics in public distribution systems
Sturm JF (1999) Using SeDuMi 1.02, a MATLAB toolbox for optimization over symmetric cones. Optim Methods Softw 11(1–4):625–653
Su X, Masoum MA, Wolfs PJ (2014) Optimal PV inverter reactive power control and real power curtailment to improve performance of unbalanced four-wire LV distribution networks. IEEE Trans Sustain Energy 5(3):967–977
Šulc P, Backhaus S, Chertkov M (2014) Optimal distributed control of reactive power via the alternating direction method of multipliers. IEEE Trans Energy Convers 29(4):968–977
Sun J, Li H, Xu Z et al (2016) Deep ADMM-Net for compressive sensing MRI. Adv Neural Inf Process Syst 29
Sun J, Chen T, Giannakis G, Yang Z (2019) Communication-efficient distributed learning via lazily aggregated quantized gradients. Adv Neural Inf Process Syst 32
Sutton RS, Barto AG (2018) Reinforcement learning: an introduction. MIT press.
Sutton RS, McAllester D, Singh S, Mansour Y (1999) Policy gradient methods for reinforcement learning with function approximation. Adv Neural Inf Process Syst 12
Sze V, Chen Y-H, Yang T-J, Emer JS (2017) Efficient processing of deep neural networks: a tutorial and survey. Proc IEEE 105(12):2295–2329

Tonkoski R, Lopes LA, El-Fouly TH (2010) Coordinated active power curtailment of grid connected PV inverters for over-voltage prevention. IEEE Trans Sustain Energy 2(2):139–147
Tran MQ, Zamzam AS, Nguyen PH, Pemen G (2021) Multi-area distribution system state estimation using decentralized physics-aware neural networks. Energies 14(11):3025
Turitsyn K, Sulc P, Backhaus S, Chertkov M (2011) Options for control of reactive power by distributed photovoltaic generators. Proc IEEE 99(6):1063–1073
Uddin MS, Kuh A, Weng Y, Ilić M (2015) Online bad data detection using kernel density estimation. In: 2015 IEEE power and energy society general meeting. IEEE, pp 1–5
van der Veen GJ, van Wingerden J-W, Fleming PA, Schol-brock AK, Verhaegen M (2013) Global data-driven modeling of wind turbines in the presence of turbulence. Control Eng Pract 21(4):441–454
Villani C et al (2009) Optimal transport: old and new, vol 338. Springer
Von Appen J, Stetz T, Braun M, Schmiegel A (2014) Local voltage control strategies for PV storage systems in distribution grids. IEEE Trans Smart Grid 5(2):1002–1009
Wang G, Giannakis GB, Chen J (2019a) Learning ReLU networks on linearly separable data: algorithm, optimality, and generalization. IEEE Trans Signal Process 67(9):2357–2370
Wang G, Giannakis GB, Chen J (2019b) Robust and scalable power system state estimation via composite optimization. IEEE Trans Smart Grid 10(6):6137–6147
Wang G, Giannakis GB, Chen J, Sun J (2019c) Distribution system state estimation: an overview of recent developments. Front Inf Technol Electron Eng 20(1):4–17
Wang G, Giannakis GB, Eldar YC (2017) Solving systems of random quadratic equations via truncated amplitude flow. IEEE Trans Inf Theory 64(2):773–794
Wang G, Giannakis GB, Saad Y, Chen J (2018a) Phase retrieval via reweighted amplitude flow. IEEE Trans Signal Process 66(11):2818–2833
Wang G, Zamzam AS, Giannakis GB, Sidiropoulos ND (2018b) Power system state estimation via feasible point pursuit: algorithms and Cramér-Rao bound. IEEE Trans Signal Process 66(6):1649–1658
Wang G, Zhu H, Giannakis GB, Sun J (2019d) Robust power system state estimation from rank-one measurements. IEEE Trans Control Netw Syst 6(4):1391–1403
Wang Y, Zou R, Liu F, Zhang L, Liu Q (2021) A review of wind speed and wind power forecasting with deep neural networks. Appl Energy 304:117766
Wang Z, Ling Q, Huang T (2016) Learning deep 0 encoders. In: Proceedings of the AAAI conference on artificial intelligence, vol 30, no 1
Weng Y, Faloutsos C, Ilić MD (2014) Data-driven topology estimation. In: 2014 IEEE international conference on smart grid communications (SmartGridComm). IEEE, pp 560–565
Weng Y, Negi R, Ilić MD (2013) Historical data-driven state estimation for electric power systems. In: 2013 IEEE international conference on smart grid communications (SmartGridComm). IEEE, pp 97–102
Wood AJ, Wollenberg BF, Sheblé GB (2013) Power generation, operation, and control. Wiley
Wu Y, Schuster M, Chen Z, Le QV, Norouzi M, Macherey W, Krikun M, Cao Y, Gao Q, Macherey K et al (2016) Google's neural machine translation system: bridging the gap between human and machine translation. arXiv preprint arXiv:1609.08144
Xiong R, Sun F, Chen Z, He H (2014) A data-driven multi-scale extended Kalman filtering based parameter and state estimation approach of lithium-ion polymer battery in electric vehicles. Appl Energy 113:463–476
Xu C, Abur A (2017) A fast and robust linear state estimator for very large scale interconnected power grids. IEEE Trans Smart Grid 9(5):4975–4982
Xu H, Domínguez-García AD, Sauer PW (2019) Optimal tap setting of voltage regulation transformers using batch reinforcement learning. IEEE Trans Power Syst 35(3):1990–2001
Yan Z, Xu Y (2018) Data-driven load frequency control for stochastic power systems: a deep reinforcement learning method with continuous action search. IEEE Trans Power Syst 34(2):1653–1656

Yang Q, Coutino M, Wang G, Giannakis GB, Leus G (2020a) Learning connectivity and higher-order interactions in radial distribution grids. In: ICASSP 2020–2020 IEEE international conference on acoustics, speech and signal processing (ICASSP). IEEE, pp 5555–5559

Yang Q, Sadeghi A, Wang G, Giannakis GB, Sun J (2020b) Deep policy gradient for reactive power control in distribution systems. In: 2020 IEEE international conference on communications, control, and computing technologies for smart grids (SmartGrid-Comm). IEEE, pp 1–6

Yang Q, Wang G, Sadeghi A, Giannakis GB, Sun J (2019) Two-timescale voltage control in distribution grids using deep reinforcement learning. IEEE Trans Smart Grid 11(3):2313–2323

Yin S, Wang G, Karimi HR (2014) Data-driven design of robust fault detection system for wind turbines. Mechatronics 24(4):298–306

Zamzam AS, Baker K (2020) Learning optimal solutions for extremely fast AC optimal power flow. In: 2020 IEEE international conference on communications, control, and computing technologies for smart grids (SmartGridComm). IEEE, pp 1–6

Zamzam AS, Fu X, Sidiropoulos ND (2019a) Data-driven learning-based optimization for distribution system state estimation. IEEE Trans Power Syst 34(6):4796–4805

Zamzam AS, Yang B, Sidiropoulos ND (2019b) Energy storage management via deep Q-networks. In: 2019 IEEE power and energy society general meeting (PESGM). IEEE, pp 1–5

Zamzam AS, Sidiropoulos ND (2020) Physics-aware neural networks for distribution system state estimation. IEEE Trans Power Syst 35(6):4347–4356

Zhang L, Kekatos V, Giannakis GB (2016) Scalable electric vehicle charging protocols. IEEE Trans Power Syst 32(2):1451–1462

Zhang L, Wang G, Giannakis GB (2019a) Distribution system state estimation via data-driven and physics-aware deep neural networks. In: 2019 IEEE data science workshop (DSW). IEEE, pp 258–262

Zhang L, Wang G, Giannakis GB (2019b) Real-time power system state estimation and forecasting via deep unrolled neural networks. IEEE Trans Signal Process 67(15):4069–4077

Zhang Y, Gatsis N, Giannakis GB (2013) Robust energy management for microgrids with high-penetration renewables. IEEE Trans Sustain Energy 4(4):944–953

Zhang Y, Madani R, Lavaei J (2017) Conic relaxations for power system state estimation with line measurements. IEEE Trans Control Netw Syst 5(3):1193–1205

Zhao J, Mili L (2018) Vulnerability of the largest normalized residual statistical test to leverage points. IEEE Trans Power Syst 33(4):4643–4646

Zhao J, Mili L, Pires RC (2018) Statistical and numerical robust state estimator for heavily loaded power systems. IEEE Trans Power Syst 33(6):6904–6914

Zhu H, Giannakis GB (2014) Power system nonlinear state estimation using distributed semidefinite programming. IEEE J Select Top Signal Process 8(6):1039–1050

Zhu H, Liu HJ (2015) Fast local voltage control under limited reactive power: optimality and stability analysis. IEEE Trans Power Syst 31(5):3794–3803

Zimmerman RD, Murillo-Sánchez CE, Thomas RJ (2010) MATPOWER: steady-state operations, planning, and analysis tools for power systems research and education. IEEE Trans Power Syst 26(1):12–19

Zonouz S, Rogers KM, Berthier R, Bobba RB, Sanders WH, Overbye TJ (2012) SCPSE: Security-oriented cyber-physical state estimation for power grid critical infrastructures. IEEE Trans Smart Grid 3(4):1790–1799